THE GREAT ENERGY [illegible]ON

THE GREAT ENERGY TRANSITION

AMERICA FROM 1876 TO 1929

DAVID E. NYE

The MIT Press
Cambridge, Massachusetts
London, England

The MIT Press
Massachusetts Institute of Technology
77 Massachusetts Avenue
Cambridge, MA 02139
mitpress.mit.edu

The MIT Press would like to thank the anonymous peer reviewers who provided comments on drafts of this book. The generous work of academic experts is essential for establishing the authority and quality of our publications. We acknowledge with gratitude the contributions of these otherwise uncredited readers.

This book was set in Bembo Book MT Pro by Westchester Publishing Services. Printed and bound in the United States of America.

Library of Congress Cataloging-in-Publication Data is available.

ISBN: 978-0-262-05212-2

10 9 8 7 6 5 4 3 2 1

EU Authorised Representative: Easy Access System Europe, Mustamäe tee 50, 10621 Tallinn, Estonia | Email: gpsr.requests@easproject.com

In memory of my sister,
Sarah Heller, 1949–2025

The changes are so rapid and radical in human society, flitting before our gaze like the fast-unfolding views of a huge panorama, or like dissolving and charming lines in the kaleidoscope, that we can scarcely get more than an outline or rightly esteem their far-reaching effects. The eye grows weary, and the brain is overpowered in contemplating these monotonous and astounding changes.
—John Wesley Edward Bowen, speech at the Atlanta Exposition, 1895

It has been called the Iron Age, the Age of Steam, the Coal Age, the Machine Age. Lately we hear, depending on the trade association involved, the Oil Age, the Electrical Age, and the Motor Age. In a year or two, the Air Age will be a well-worn phrase.
—Stuart Chase, *Men and Machines*, 1929

CONTENTS

INTRODUCTION

There were so many simultaneous transformations between 1876 and 1929 that Americans struggled to understand them. They lacked both a language and a theory adequate to the task. They saw everywhere the social and economic effects of a vast increase in energy, but they thought about it in terms of force, and they lacked the concept of an energy transition. Nor did they yet think in terms of the concept of technology. Instead, they struggled to speak of change in terms of "the practical arts" or "applied science." American historians inherited this semantic inadequacy, and they did not conceive what occurred in the United States between 1876 and 1929 as an energy transition and the beginning of the Anthropocene. Instead, they defined these years as a second industrial revolution, the era of reform, the Gilded Age, or the Progressive Era, followed by the Great War, and the "Roaring Twenties."

This book presents a more unified understanding of these years by reconceiving them as a great energy transition, from horses to automobiles, from steam to electricity, from burning coal and wood to gas and oil, and from muscle power to mechanical power. This energy transition enabled the transformations of that era, including the consolidation of a national market, enormous urban growth, increased immigration, new forms of mass production, higher wages, new forms of transportation, enormous engineering projects such as dams and irrigation systems, and new styles of architecture. Those alive at the time knew they had witnessed the emergence of department stores, skyscrapers, accelerated communication, the concentration of manufacturing, the increased availability of consumer goods, and a flood of immigrants. They celebrated high-energy inventions that transformed everyday life, such as the electric light, cinema, automobiles, radio, and airplanes. But they perceived these many aspects of the energy transition as a myriad of disconnected events. Unable to discern or conceptualize an underlying cause, they often declared that the collective transformations were "kaleidoscopic." This metaphor suggested rapid, ungraspable, but largely beneficent transformations based on practical applications of scientific discoveries. And because scientific laws were immutable, it seemed that the transformation of society was inevitable.

An energy transition is not merely kaleidoscopic, nor is it merely a shift from one kind of fuel to another, nor just a matter of increasing horsepower. The psychological dimension of the change is also a central if more elusive part of the transformation. The Americans of 1929 did not just command more power; they thought about themselves and saw the world in a different way. To understand this aspect of an energy transition, it is helpful to think in terms of a changing lifeworld. Edmund Husserl developed this concept, which was later taken up by Jürgen Habermas and many others, to describe what a person experiences as the ever-present characteristics of daily life, or the things taken for granted as being always already there.[1] This is not an idiosyncratic private perception of the world, but one widely shared as background knowledge that provides the horizon of a common understanding. The sense of the lifeworld frames communication and circumscribes what can be considered reasonable. To take just one example, in the lifeworld of 1875 it was axiomatic that the sound of a human voice could only be heard if a person was present. It had always been so. Speech could not be separated from human presence. In contrast, in the lifeworld of 1929, a child took the phonograph and the radio for granted. After 1875 it had become possible to send and receive the sound of a voice through the air or over a wire. By 1929, the ability to preserve, replay, and manipulate a voice had been naturalized. One can make a long list of such differences but taken together they are far more than a sum of the parts. The many elements of a lifeworld collectively establish a perceptual structure that shapes the sense of what is natural and real. The world one is born into has distinct shapes, textures, rhythms, and energies, and together they form the horizon of experiences that seem familiar and comfortable. Habermas suggested that we "can think of the lifeworld as represented by a culturally transmitted and linguistically organized stock of interpretive patterns."[2] These patterns can change radically, however. Many things that had been axiomatically defined as impossible in 1875 were normal in 1929. And when there are many such alterations, the lifeworld as a whole is altered. The great energy transition entailed a fundamental change in what one took for granted about the world, making necessary new interpretive patterns.

The changes in the American lifeworld were enormous between 1876 and 1929. In 1876 people mostly relied on muscle power to move about. They owned millions of horses, mules, and oxen, not just on farms but in cities as well. By 1929 new technologies had displaced this animal power, in the form of streetcars, bicycles, automobiles, tractors, trucks, and buses. In 1876, speakers had to rely on their vocal cords to speak to an audience. In 1929, the microphone, radio, and talking pictures amplified messages to millions of listeners. In 1876, the telegraph was the only electrified media, and no one had heard recorded sound, made a telephone call, or seen a movie. Few had heard the voice of an American president. They had never seen an electric light, much less a spectacular flashing sign. In 1876 the city was horizontal, not vertical. Few buildings were more than four stories high, and no one knew the word "skyscraper." There were just a few elevators and no escalators, no subway systems, no elevated train lines, and no amusement parks. There were no microphones and loudspeakers to amplify a singer's voice, no film stars, no jazz music, and no professional sports stadiums. There were no automats, no vending machines, and no ice cream cones. Most factories were small and inefficient, and none had assembly lines. In 1876 few had seen a bicycle, and the electric streetcar, the automobile, and the heavier than air flying machines did not yet exist. No one lived in a home that was electrified and few houses had flush toilets. There were no gasoline-powered machines such as lawnmowers, and no electric irons, ovens, or toasters. These contrasts suggest the differences between the lifeworlds of 1875 and 1929.

The great energy transition also included the transformation and networking of cities, as the urban population mushroomed from 9.9 to 69 million, shifting a majority of Americans away from rural life. This urban concentration was only possible because agriculture industrialized. Farmers became more productive, using manufactured fertilizers, hybrid seed, and high-energy machines. The farming population declined but food production increased. Nor could the cities have flourished without a vast infrastructure of roads, mass transit, sewers, water supplies, gas pipes, electrical wires, elevators, refrigerators, fire alarms, signaling systems, gasoline stations, and much else powered by steam engines, electricity, gasoline motors, or natural gas. The management and control of these systems demanded technical expertise, political engagement, and many reforms. New energy systems also made possible a national market dominated by corporations that did not exist in 1875, the redefinition of blue-collar work, the creation of millions of white-collar jobs, and the engineering of nature into an improved "second creation." New forms of communication and transport increased physical mobility and detachment from the local in favor of the national, the redefinition of time amid the acceleration of events, the commodification of both sound and image, and the invention of media events. These changes encouraged admiration for the technological sublime and became inseparable from the aesthetics of modernism. These examples suggest the comprehensive effects of America's largest energy transition.

However, like those alive at the time, most historians have not paid much attention to the great energy transformation, which was also the dawn of the Anthropocene. They focused on the aftermath of the Civil War when leaders sought to reunite North and South, first through Reconstruction and later at the cost of racial segregation. They examined unions and struggles to organize labor. They analyzed immigration, the fight for women's suffrage, and discrimination against ethnic, racial, and LGBTQ groups. Even Howard Mumford Jones's *The Age of Energy* scarcely discussed and did not even index such topics as assembly lines, dams, electricity, energy, engineering, irrigation, mass production, technology, or the telephone. Nor did Jones refer to Thomas Edison, the Wright brothers, or Henry Ford. More recently, such topics and individuals may be mentioned, but there is no recognition that a great energy transition underpinned the myriad transformations of that era. Instead, there were studies of the oil, gas, and coal industries, of local utilities, of electrical equipment manufacturers, and of the competition within the energy sector. Economic historians studied the battles between gas and electrical companies for lighting and heating markets; between coal and oil for home heating; between waterpower, steam engines, and electric drive in the industrial market; between electricity and gasoline in transportation; and other clashes in the marketplace. This competition distracted their attention from the larger fact that the total energy available was increasing rapidly and that competition drove down prices and hastened the energy transition as a whole.

The energy transition was also overlooked in part because the concept of energy itself was still new in the late nineteenth century. Most Americans did not think in terms of either "energy" or "technology." As Eric Schatzberg explains in his landmark history of the concept "technology," "at the end of the nineteenth century, no English-language term seemed suitable to describe the momentous changes associated with the Second Industrial Revolution," and many authors simply spoke of "the machine" or "applied science." They suffered from "what Leo Marx has called a 'semantic void,'" due to "the absence of adequate concepts to describe the new industrial era."[3] The new energy systems were social constructions, not inevitabilities, but Americans "found it difficult to grasp the nature of the changes they were experiencing."[4]

After circa 1850 there was a shift among scientists from the concept of "force" to that of "energy." But for most Americans "force" remained the central term in explaining movement, whether the gravitational movement of planets, the muscular power of horses, or the mechanical action of machines. Lord Kelvin and other scientists realized that "force" was a metaphor more than a precise concept, and they replaced it with the laws of thermodynamics, which entailed a new understanding of the universe in which "energy" was a central term. But most Americans continued to think in terms of force until the 1920s. As Ronald Martin has trenchantly explained, they also believed in the "conservation of force," an idea at odds with the concept of entropy, the second law of thermodynamics.[5] Why does this matter? Because Americans believed that the conservation of force meant that progress was inevitable and cost-free. It seemed that when society gained control of greater forces, it automatically advanced. The spokespersons for this view included the popular English author Herbert Spencer[6] and many Americans, such as the industrialist Andrew Carnegie, the clergyman Henry Ward Beecher, the prominent sociologist Simon Patten, and a former president of the American Society of Mechanical Engineers and professor at Cornell, Robert Thurston. He was certain that continual increases in power in factories and transportation were the prime movers of social and economic progress. He declared that "from the perfection of [James] Watt's steam engine and its application to mills and factories, and to steamboats and railroads, wealth has accumulated with a steady increasing rate of accumulation."[7] Thurston concluded that because of their greater control of force, Americans had higher wages, more leisure time, and improved health. Thurston and most of his generation embraced this deterministic theory of progress.[8]

The great energy transition affected every social group to some degree, cutting across racial and ethnic differences. The civic culture of 1876 was based on a far different lifeworld than that of 1929. Contrast a small crowd in 1876 watching an amateur baseball game in a meadow with an urban ballpark filled with 50,000 fans watching professional athletes in 1929. Contrast a horsecar with a subway train. Contrast a small crowd listening to a speaker's unaided voice in 1875 with an orator broadcasting from a radio studio to millions in 1929. As such examples suggest, the lifeworld changed profoundly. Regionalism attenuated, as the nation became a single market, drawn together by faster transportation, corporate consolidation, mass marketing, and new forms of communication. Regionalism was being replaced by kaleidoscopic celebrations of distinctive locations that together represented the nation.

This argument does not treat technology as a deterministic force, in which inventions inexorably transform society. Instead, like most historians of technology, I argue that inventions and their adoption are socially constructed processes, whose outcomes are not predetermined and at times unexpected.[9] Thomas Edison thought the phonograph would be widely adopted in business and education, but the public preferred to use it as a music box. The first experimenters on radio saw it as a wireless form of point-to-point communication that could replace the telegraph, but it was more widely used as a broadcast medium. The uses of new technologies were not imposed by inventors but were social constructions co-produced by entrepreneurs and consumers. The many-faceted transformation of American society between 1876 and 1929 was not imposed but negotiated.

Take the change in diet as an example. Between 1870 and 1929 the per capita consumption of beef and pork fell from 193 to 110 pounds per year. When consumers were able to vary their diet, they annually chose to buy an additional 120 pounds of fruit and 100 pounds of vegetables instead of meat. This choice became possible because the perfection of canning gave a long shelf life to peas, carrots, corn, green beans, blueberries, and many other foods, while the development of refrigerated railway cars and steamships brought to eastern markets fresh lettuce, melons, bananas, tomatoes, citrus fruits, and other produce from Central America, the South, Texas, and California. Both canning and refrigerated transport were also adopted by meat producers. Nevertheless, consumer preferences drove a change in demand. Meat producers advertised vigorously, but their sales per capita fell 43 percent.[10] Fortunately for them, population growth made up for this loss.

Economic historian Robert Gordon has documented that the years between 1870 and 1929 were a period of unusually rapid growth, when both productivity and the quality of everyday life increased at a higher rate than during the decades before or after. The improvements were measured in higher wages, shorter working hours, longer life expectancy, reduced infant mortality, and the widespread installation of running water, flush toilets, telephones, and electricity. Entirely new kinds of experience also emerged with the inventions of the phonograph, automobile, airplane, and film. Gordon recognizes that statistics do not entirely capture the value of these improvements. Statistics do not evoke a lifeworld. The greater speed of electric streetcars compared with horsecars can be quantified, but it is also important to realize that horse manure disappeared from city streets, which reduced the number of flies and foul odors. Electric streetcars could be warmed in winter, but the value of such comfort is not easily quantified. Likewise, many improvements were not captured by statistics, including the value of "the brightness, convenience, and safety of electric light"; "the reduced risk of food contamination made possible by refrigeration"; "instant communication" through the telephone; experiences provided by the phonograph, radio, and motion pictures; and having clean running water inside the home.[11] Such things were rare or unknown in 1876. The fabric of daily life had been transformed, weakening rural isolation, and linking homes, businesses, and civic institutions into networks of water, sewage, electricity, gas, transportation, commerce, and media. To be sure, class and racial differences sharpened, and women still faced discrimination in the workplace, even if they did win the right to vote. Nevertheless, an overall comparison of 1876 with 1929 discloses a radical change in the lifeworld.[12]

The great energy transition also upended the business hierarchy and remade the landscape. These changes were not sudden but incremental, and they required roughly half a century.[13] As Vaclav Smil notes, "shifts to new sources of primary energy have been gradual, prolonged affairs."[14] Energy transitions take fifty years or more, and they start slowly and then accelerate after about two decades.[15] During a transition, older technologies either decline or are repurposed. Buggies, kerosene lamps, and bed warmers filled with glowing coals gradually disappeared, but water wheels and steam engines could be converted to generate electricity.

An energy transition has five phases: invention, trials, expansion, momentum, and maturity. In a landmark study of the electrical industry, *Networks of Power*, Thomas Hughes laid out the phases of a transition, which culminates when it achieves "technological momentum," a concept that Hughes also applied to other large-scale systems, such as the railroads. The first phase of invention arouses interest through public demonstrations at a few locations, but it has little immediate effect on the economy. In the second phase a new energy system begins to be showcased at a few locations, leading to a third phase of growth and new applications. The system expands as solutions are found for bottlenecks that Hughes calls "reverse salients." As he explains, "Networked systems consisting of a number of heterogeneous components or firms evolve like a shifting front. The components and firms that fall behind, I name 'reverse salients.' . . . Reverse salients in networked systems need to be corrected in order for the systems to continue to evolve. An example of a reverse salient in early electric power systems was the absence of a satisfactory motor for alternating current systems."[16] As such difficulties are overcome, the industry can expand and lower unit costs. Only then does an energy transition reach the crucial fourth phase of "technological momentum," when the new energy becomes the preferred form of light, heat, or power. Electricity reached technological momentum by circa 1910, at almost the same time as motorized transportation. Nevertheless, in 1915 only 15 percent of US homeowners had electricity or automobiles, and only 20 percent of factory power was electric.[17] To deal with the legal, organizational, and logistical problems that arise during further growth in the fifth, mature phase requires a new kind of management that includes financiers and consulting engineers.[18] The great energy transition was complete by 1929, and energy use accelerated in the 1930s. Electrified streets, business districts, factories, and homes had replaced older energy technologies of light and power in urban areas, and they were moving into the hinterland. Likewise, in the 1920s automobiles, trucks, and buses reached the mature stage.

The concept of technological momentum suggests why an energy transition requires a long time. A new system is flexible in the initial phases but gradually becomes more rigid as technical specifications are standardized and widely implemented. Once voltage is standardized, for example, it is prohibitively expensive to alter it, as everything plugged into an electrical system has been configured to use that voltage. Development of a bureaucracy to run the system also makes the system more rigid.[19] Once a socio-technical system achieves momentum, it can grow rapidly but will resist radical changes. It has been woven into society and become a "natural" part of the lifeworld. People become dependent on it, and soon they can scarcely imagine their lives without it. The same applies to the individual elements of an energy system such as the electric toaster, doorbell, wall switch, or automobile.

The five phases of the great energy transition are an essential background, but this book is not organized into five chapters based on these five phases. Such an organization does not explain how Americans experienced a major transformation in their lifeworld. Instead, part I, "Transition," consists of four chapters that examine the great transition, primarily before it reached technological maturity in the 1920s. Chapter 1 examines how Americans tried to make sense of rapid change. They did not perceive the energy transition but rather its "kaleidoscopic" effects and experiences. Chapter 2 examines the energy transition in cities, employment, and transportation. Chapter 3 turns to the engineers, who grew from a small cohort in 1870 to several hundred thousand in 1929. They were celebrated as heroes, as they spearheaded construction of the new energy infrastructure of paved roads, mines, dams, pipelines, refineries, power plants, electric transmission systems, and skyscrapers. Chapter 4 then looks at the disintegration of the lifeworld of circa 1870, replaced by a more malleable conception of space, a vastly longer stretch of geological time, dynamic forms of leisure, and a new terminology of identity based on energy metaphors. The new lifeworld fragmented space, accelerated experience, commodified images and sounds, and problematized identity. Together, the four chapters of part I show how ordinary people responded to these transformations that were attractive, upsetting, and confusing, all at the same time.

Part II, "Reactions," focuses on reforms and resistance to change. Chapter 5 examines both the progressive reforms that responded to the surging immigration, urbanization, and technical change and the reactionary reforms developed to segregate, Americanize, or exclude those who did not "fit" the kaleidoscopic imaginary. Chapter 6 examines nostalgic reactions against the new lifeworld, as many Americans identified with earlier times. Chapter 7 explores how Americans redefined traditional individualism and celebrated heroes of the energy transition.

Part III, "Momentum," deals primarily with the later years of the energy transition as it achieved technological momentum. Chapter 8 surveys how the new energy technologies were used to reshape the economic hierarchy, which by the 1920s was dominated by corporations that had not existed in 1876. Their products, such as the phonograph, motion pictures, automobiles, electrical appliances, and radio, were unknown in 1876 but had become central parts of the new high-energy lifeworld. Chapter 9 investigates how the media event emerged, making simultaneous nationwide experiences possible, between the dedication of the first transcontinental railroad in 1869 and the jubilee of the electric light in 1929. Chapter 10 discusses the great energy transition in its mature stage during the 1920s, when the kaleidoscopic metaphor declined and was replaced by the more deterministic concept of "the machine age." Even so, Americans still lacked the conceptual tools needed to comprehend the great energy transition.

The epilogue looks briefly at the United States of 2025, when renewable energies had achieved technological momentum but were stoutly resisted by a Republican administration that would prefer to remain in the lifeworld of a century ago.

I

TRANSITION

1

A KALEIDOSCOPIC NATION

The kaleidoscope was invented in 1816 by a Scottish scientist, Sir David Brewster, who received a British patent for it in 1817.[1] It soon became popular in the United States. Look through a kaleidoscope, and you will see at the end of the tube shards of colored glass that are reflected by mirrors to create symmetrical patterns. These change when the tube is rotated, to produce what Helen Groth describes as "a sense of perpetual transformation . . . that tested the limits of verisimilitude and reflected the existential flux of modern life." In the compound kaleidoscope the small shards of colored glass are removed, and one sees instead the mirrored repetition of fragmented forms of the surroundings. As Brewster explained, this can make anything kaleidoscopic: "The furniture of a room, books and papers lying on a table, pictures on the wall, a blazing fire, the moving branches and foliage of trees and shrubs, bunches of flowers, horses and cattle in a park, carriages in motion, the currents of a river, waterfalls, moving insects, the sun shining through clouds or trees, and, in short, every object in nature may be introduced by the aid of the lens into the figures created by the instrument."[2] This playful fragmentation and multiplication of vision became a popular metaphor for the rapid transformations of modernity. Groth concluded that "kaleidoscopic metaphors oscillate, evoking either positive associations with a self-consciously cosmopolitan gaze open to diversity or equally suggestive negative associations with the infinite and alienating interchangeability" of industrial-scale production. In either case, there is "a relentless exchange of images that never quite come to rest in time and space."[3]

After the Civil War, Americans began to use the kaleidoscopic metaphor for their society, which seemed endlessly to generate new patterns by rearranging the materials of everyday life. In American publications the use of the word "kaleidoscope" doubled between the 1860s and the 1870s, quadrupled to a peak between 1900 until 1920, and then gradually declined.[4] For half a century, the word served as a popular metaphor for a changing society. A San Francisco newspaper declared in 1875 that "society may be compared to one of those brilliant forms which the kaleidoscope produces; men's natures, like the gay bits of tinsel and glass, are always the same, but their combinations are infinite in variety; the poetry of society catches the shifting picture before it dissolves into another scene."[5] The varied contents that were rearranged did not include those who deviated from middle-class norms. The kaleidoscopic metaphor was used to celebrate a limited range of social and ethnic differences, while touting the economic and technological progress of the United States. The metaphor was not used to describe poverty, crime, or the devastation of natural disasters. Rather, it referred to pleasing multiplicity and increasing complexity. It could even be used to describe the diversity of government functions. As Calvin Coolidge wrote in his 1927 Budget Message to Congress, "In our great and growing government changes are kaleidoscopic and new needs and imperative new demands arise overnight."[6]

To grasp the kaleidoscopic metaphor's context, recall that during the preceding two decades the Civil War had torn the nation apart and Reconstruction had forcibly reunited North and South. Nationalism defeated regionalism, industrialization overwhelmed agrarianism, and free labor triumphed over slavery. Yet racism remained endemic in the United States and white men retained privileges not extended to women, formerly enslaved people, Native Americans, or Asian and Hispanic immigrants. When a post–Civil War imaginary emerged, including the kaleidoscope, it did not challenge this inequality. The kaleidoscopic metaphor provided a way to celebrate a reunified country without acknowledging hierarchies or racial differences. The kaleidoscope suggested the restless multiplicity of a regional nation being consolidated. The metaphor also proved useful in celebrations of technological triumphs such as dams, bridges, and skyscrapers. The nation was becoming multifaceted. Corporations were mushrooming to dominate whole industries, and the variety of jobs increased exponentially. Advertising grew from simple, short texts filling a few lines in a newspaper to become an industry, and public relations emerged at almost the same time. Millions of immigrants poured into the country. The buildings, the lights, and the sights of the city literally changed. The lifeworld was in transformation, and the metaphor of a kaleidoscope suggested this unceasing change and variety.

Between 1880 and 1929 the factory work week decreased from sixty-one to forty-eight hours.[7] Labor unions fought to obtain this additional leisure time, and there were many confrontations between management and workers.[8] But industries also discovered that time reductions were advantageous. When a Philadelphia company voluntarily reduced hours from 57.5 a week to 52.5, it told workers about the deleterious effects of fatigue and asserted that "if they worked shorter hours and had a greater rest period they could do as much work in a short time as they did in the longer time."[9] This proved to be true, and further reductions followed. Strikes often proved necessary to pressure management, but reducing hours while increasing pay was possible because efficiency improved. Jobs became more efficient partly by eliminating breaks, horseplay, and amusements, but the largest gains were due to adoption of new work processes and labor-saving technologies.

Increasingly, work and play were defined as separate activities. The new leisure time was filled with new amusements, including vaudeville, dance halls, bicycling, roller skating, amusement parks, penny arcades, movie houses, cabarets, and professional sports. The crowds who frequented these new venues were diverse. They included not only numerous blue-collar workers but also a rapidly increasing white-collar class. As David Nasaw noted in his study of popular urban amusements, there were only 5,000 typists and stenographers in 1880, but thirty years later they numbered 300,000. The clerical workforce, many of them women, increased to 1.7 million. They were concentrated in the largest cities, and they "were the most avid consumers of the commercial pleasures."[10]

An equally important change in urban life was the influx of Europeans. Immigration was not new, but its intensity increased. In every decade from the 1870s until World War I, the numbers of immigrants grew, from 5 million in the 1880s to 8.7 million in the 1900s. As early as 1890, immigrants and their children were the majority in the fifteen largest cities. Less than one in five residents of Milwaukee, New York, or Chicago was native-born of native parents; less than one in three were in Detroit, San Francisco, Buffalo, Cleveland, St. Paul, St. Louis, Cincinnati, Pittsburgh, Boston, or New Orleans. Only in Baltimore and Washington, DC, were almost half of the population native-born of native parents.[11] Cities developed distinctive ethnic neighborhoods that included churches, fraternal associations, grocers, barbershops, saloons, and other institutions catering to each ethnic group. The kaleidoscope metaphor could include these diverse ethnic communities within the dominant white culture.

The variety of immigrant communities became a common theme in descriptions of American cities. A visitor to California exclaimed, "Street life in San Francisco is a kaleidoscope that is never at rest. There is nothing like it on the continent."[12] He went on to extoll the flowers, the fruit stands, the lively shops, and the tides of people, including "strolling players" offering entertainment under an open sky. A visitor to Pittsburgh a year later reported to readers of *Harper's New Monthly Magazine* that "the City of Smoke" offered "a hundred objects of interest that are beyond the scope of this article" including the "humming hives of whirling spindles" in its mills, "a firmament of lights flashing on the swift water of three rivers," and impressive iron bridges. There was also the "great army of ten thousand men" who thronged its "gas-lit thoroughfares" each evening, "seeking amusement." Overall, "to the intelligent observer, Pittsburg is a great kaleidoscope, showing new attractions at every turn."[13] Likewise, a guidebook praised the city of Washington, DC, as being "like a kaleidoscope."[14]

Such guidebooks were published because more Americans were taking vacations. While some resorts had emerged before the Civil War, the era of mass tourism might be dated from 1872 when the English touring entrepreneur Thomas Cook opened a New York office, soon to be followed by branches in other cities.[15] They sold package tours that included travel, accommodation, and information about the sights at each stop, typically emphasizing one impressive destination, such as Niagara Falls, Manhattan, or a resort hotel on one of the Great Lakes. American travel agencies proliferated and offered similar services. They collaborated with railroads, which were encouraging the creation of hotels, resorts, and national parks along their lines. The Northern Pacific lobbied Congress to create the first national park at Yellowstone and then built a station there. The Great Northern Railroad ran its tracks by the entrance to the new Glacier National Park. The Santa Fe built a spur line to the Grand Canyon and put a magnificent hotel on the rim. Railroads also encouraged tourism to cities along their lines. Competition kept prices down, and more Americans traveled long distances on vacations. Camping, hiking in the mountains, and sea bathing became popular, and the perception spread that the nation had many attractive destinations.

As Nina Silber summarized, in the later nineteenth century tourism became "a mass industry predicated on transporting herds of people through strange and unusual scenes." The American West was prized for its snow-capped mountains, vast canyons, and sublime novelties. The South was romanticized as an anti-modern refuge from industrialization that "became less the subject of sociological or political scrutiny" than it had been during the sectional crisis, and instead was a "land filled with unique sites and attractions" such as ruined plantations, cotton fields, and picturesque towns, "that stood in stark contrast to the commodified uniformity of northern society."[16] Some travelers wintered in Asheville, North Carolina, or Saint Augustine, Florida. Others preferred larger cities. Urban landscapes were being celebrated in grand panoramic photographs, notably Eadweard Muybridge's eleven plates depicting San Francisco in 1877 and similar images of other emerging metropolises. These drew inspiration from painted panoramas that had emerged in Britain in the late eighteenth century and spread throughout Europe and the United States.[17] Photographers presented the new cities as idealized landscapes. The *Chicago Art Portfolio*, published in 1894, depicted public buildings, monuments, boulevards, clubs, hotels, churches, parks, homes of the wealthy, and portraits of industrialists, mayors, and other leaders. Similar books depicted San Francisco, New York, and other cities. In each, panoramic views provided an overview followed by images of an orderly urban world.[18] The reader who flipped through these books saw a kaleidoscope of views that omitted slums, stockyards, poorly constructed buildings, smoke pollution, beggars, or anything unsightly. They intimated that cities were rapidly evolving toward an almost utopian perfection.

The artist Joseph Pennell (1857–1926) celebrated cityscapes and spectacular projects. His drawings, etchings, and lithographs were reproduced in books and national magazines, and sold to collectors. Pennell seldom depicted individuals and focused on constructed landscapes, such as his panoramic view of the New York skyline in *Sunset from Williamsburg Bridge* (figure 1.1). He was more likely to foreground the steel beams of skyscrapers under construction than the workers involved. He interpreted America in terms of dynamic engineered progress, illustrated by factories, railway viaducts, the new Woolworth Building, or the construction of the Panama Canal. His work exemplified the celebratory, kaleidoscopic depiction of the nation. (See figures 3.1–3.4.)

Figure 1.1
Joseph Pennell, *Sunset from Williamsburg Bridge*, 1915, etching and drypoint. Courtesy of the Library of Congress, Prints and Photographs Division.

Photographers, armed with the new dry plates, also began to explore. The old wet plates had to be prepared in a darkroom immediately before use, and it took half an hour before an exposure could be made. Freed of this constraint after circa 1880, a photographer could leave the darkroom behind and roam the streets with a camera and dry plates ready for immediate exposure. Some made stereographic images of landmarks; others such as Jacob Riis exposed slum conditions or documented factory conditions. An artistic movement also emerged. Beginning in 1890, Alfred Stieglitz photographed "the streets of lower New York," including images made in the "rain, fog, and snow, and at night." As Alan Trachtenberg summarized, the pictorialism that Stieglitz championed saw photography in terms "of styles, of changing looks and aesthetic intentions isolated from social practices, cultural patterns, and institutional forms."[19] Such artistic detachment was another example of kaleidoscopic vision, which celebrated views that seemed disconnected from politics or social inequality.

Many sites explored by artistic photographers also appeared on postcards, which came into widespread use. Plain postcards with one side for a message and the other side for the address were approved by Congress in

1872, and the first penny card was sold in 1873. Before then mailing a printed card cost the same as a regular letter. The picture postcard that devoted one side entirely to an image became popular during the Columbian Exposition in Chicago. Popular tourist sites and landmark buildings were soon represented on millions of cards, and after 1898 these could also be mailed for a penny. By 1908, seven hundred million postcards annually went through the mail, and millions more were never sent but kept as souvenirs.[20] Postcard collections literally represented the nation as a series of disconnected views.

The literary expression of kaleidoscopic vision came in the form of local color writing in the realist mode, which flourished after circa 1890. The critic James M. Cox argued that after the Civil War regions were subjugated to the national state and lost "political possibility," and instead developed the "imaginative space" of place as a "literary possibility."[21] It might seem that focusing on the distinctive and the local signaled a failure of cohesion. But John Dewey understood that the defeat of the Confederacy had undermined the power of the states. The stronger "the legal unity, the more intense becomes the local life. The defeat of succession diversified the South even more than the North."[22] Amy Kaplan observed of the local color movement that the "decentralization of literature contributes to solidifying national centrality by reimagining a distended industrial national as an extended clan sharing a 'common inheritance' in its imagined rural origins."[23] Local color fiction was often framed as a contrast to urban life, as in Hamlin Garland's *Main-Travelled Roads*, where midwestern scenes implicitly contrasted with the East. Such works were written more for urban and educated readers than for the colorful local people they described. The attractions of the local stood in contrast to what George Sharp called, in 1915, "the superficiality of urban life." He felt that "in looking at the kaleidoscopic movements of the men and women who make up a big city, we focus little attention on individuals. . . . The city is the organism of which the citizen is but a cell. . . . The city man loses not only the picturesqueness of individuality, but all charm of personality. He becomes a mere shuttle, moving, by force of necessity, to and from, between home and workshop, store, or office."[24]

If many authors took readers from the city to farms, small towns, or picturesque islands, other writers explored distinctive urban neighborhoods. New York City proved especially suited to the kaleidoscopic metaphor, and postcards depicted its neighborhoods, iconic buildings, bridges, and famous sites, including Times Square, the Statue of Liberty, public library, railroad stations, bridges, and skyline. One popular guide described the streets and impressive buildings of Manhattan and then declared, “Up and down this vista roars and streams an ocean-tide of travel and traffic, and the eye can find food for continual interest in its changing kaleidoscope.”[25] *New York Illustrated* (1881) focused on Broadway, which was “exceedingly brilliant and picturesque” (figure 1.2):

> As far as the eye can reach it gathers in a range of business palaces, representing every variety of taste, style, and beauty, while between them and on the sidewalk is an ever-changing scene in which light, color, and motion combine to create a charm that never tires. There is a fascination even in the throng of vehicles, the faces in the omnibuses and private carriages, the gay turnouts, and handsome equipages; and in the strange commingling of people passing to and fro, representing every State and country, every style of dress from that of the Oriental to the last fashion of the Anglo-Saxon, there is a magnetic attraction that compels the stranger to linger and enjoy the kaleidoscopic scene. For three miles the change is continual, the continuity of effect unbroken.[26]

Some observers generalized further, seeing the kaleidoscope as a comprehensive metaphor. Herbert Hoover said in 1929 that “new ideas must

Figure 1.2
“Broadway, South from the Post-Office,” *New York Illustrated* (New York: Appleton & Co., 1881), p. 36.

be recruited from the kaleidoscope of a great shifting mass of humanity," suggesting that diversity stimulated innovative thoughts.[27] The African American reverend John Wesley Edward Bowen adopted the kaleidoscopic metaphor in an address on "Negro Day" at the 1895 Atlanta Exposition. He used it to suggest the difficulty of comprehending the rapid transformations of the age. "The changes are so rapid and radical in human society, flitting before our gaze like the fast-unfolding views of a huge panorama, or like dissolving and charming lines in the kaleidoscope, that we can scarcely get more than an outline or rightly esteem their far-reaching effects. The eye grows weary, and the brain is overpowered in contemplating these monotonous and astounding changes."[28]

Between 1876 and 1930 the lifeworld was shifting from a rural to an urban society; from a nation that relied on steam power to one based on gasoline and electricity; from a transport system based on horses and railroads to electric streetcars, trucks, automobiles, and airplanes; from small shops to giant emporiums; from family-owned businesses to modern corporations; from individual inventors to research teams in laboratories; from labor-intensive artisanal production to capital-intensive assembly lines; from local products to national brands; and from amateur athletics to professional sports, to make a partial list. The economic turbulence of the energy transition suggests why the metaphor of the kaleidoscope was appealing. It not only avoided reference to the subordination of the South after the Civil War, but suggested a benign interpretation of disquieting trends like the instability of employment, sudden industrial mergers, and frantic urban growth. The country was described as a tapestry of contrasts, a shifting kaleidoscope of progress. The large triumphed over the small, the national over the local, the rapid over the slow, and the mass-produced over the custom-made. These changes could also be thought of in terms of inventions, notably the telephone, phonograph, electric light, assembly line, Bessemer steel, the microphone, electric motor, elevator, streetcar, motion picture, airplane, and automobile. Americans used these technologies to transform their lifeworld as they pushed buildings into the sky, burrowed tunnels beneath the city, constructed elevated railways, illuminated vast interiors, and created new venues for work and leisure (figure 1.3).

In theory, these changes benefited all Americans, but in practice they were unevenly distributed. African Americans were making great strides, but they also met concerted opposition. In his Atlanta address, Bowen spoke

Fulton Street, looking toward Brooklyn Ferry.

Figure 1.3
"Fulton Street, Looking Toward Brooklyn Ferry," *New York Illustrated* (New York: Appleton & Co., 1881), 15.

of the advances made by Black Americans since the abolition of slavery, but he was careful not to offend the white leaders who listened to his address. Like Booker T. Washington, who spoke at the same exposition, Bowen depicted a future where material progress went hand in hand with racial harmony. He rhapsodized that "a new Negro has come upon the stage of action. As you enter the main entrance of the Negro building you will observe the statue of a Negro with broken manacles upon his wrists. This statue was born in the fruitful brain of a Negro, Mr. Hill of Washington. His frame is muscular and powerful; his eye is fixed upon his broken but hanging chain; his brow is knit in deep thought. This is the new Negro. What is he doing? He is thinking! And by the power of thought he will think off those chains and have both hands free to help you to build this country and make a grand destiny of himself."[29]

Bowen's image of a mesmerizing, kaleidoscopic sequence of events proved to be an accurate description of white society, but his expectations for racial improvement were too optimistic. In the generation after the Civil War some Black Americans acquired education and advanced themselves. But they faced continual opposition, not least in Georgia where Bowen made his address. The change in racial relations that he anticipated proved illusory. In the 1890s the Supreme Court cemented the legitimacy of segregation and ensured that the African American lifeworld would not be the same as that for white Americans.[30] Jim Crow laws dictated separate, poorly funded schools for Black children throughout the South. Lynchings and intimidation forced Black Americans either to flee the South or to accept their legal definition as second-class citizens. Likewise, immigration from China and Japan was curtailed, and Native Americans struggled to maintain their traditions against continual pressure to assimilate to white society. Kaleidoscopic America celebrated varieties of whiteness that included European immigrants, but it assigned other races marginal or subservient roles. This was particularly so in the former Confederate states, which were less urban and received fewer immigrants than the North.

Given the limitations of kaleidoscopic vision, some authors argued that it was little more than meaningless, shifting patterns. Fanny Kemble compared looking through a kaleidoscope to feeble thinking. To a friend who feared her mental powers were diminishing, she replied, "Your thinking powers, which appear to you so miserably deteriorated, are now not at all below the average of those of most of your fellow females, to whom what you call cogitating is a process unknown, and whose brains, like a child's kaleidoscope, are filled with odds and ends, more or less rubbishy, by dint of shaking, and the simplest possible adjustment of means producing accidental arrangements, keeps them amused and lively, by a succession of tolerably pretty and symmetrical patterns, without any real value, connection, or significance."[31] The carousel was another common metaphor for kaleidoscopic distraction, as was the Ferris wheel. In both, the individual was thrown into motion and saw surroundings as a whirling sea of change. It was impossible to fix the eye on any external point for more than a few seconds. There was no progress implied in this motion, as the machines revolved endlessly in the same place.

Mark Twain adopted many new words in his writing, and he used "kaleidoscopic" in *The Gilded Age*, along with such new words such as "six-shooter," "groggery," "dissecting room," and "buffalo hunt."[32] Twain later observed that the kaleidoscope was an instrument of repetition that masqueraded as an engine of change. "There is no such thing as a new idea. It is impossible. We simply take a lot of old ideas and put them into a sort of mental kaleidoscope. We give them a turn, and they make new and curious combinations. We keep on turning and making new combinations indefinitely; but they are the same old pieces of colored glass that have been in use through all the ages."[33] In keeping with this view, in "What Is Man?" Twain questioned whether human beings had free will or were the masters of their fate. But he nevertheless admired many "self-made men" of the time, notably General Grant, Andrew Carnegie, and H. H. Rogers of Standard Oil. Twain was ambivalent.

Willa Cather was also ambivalent in 1894 when she reviewed a theatrical performance in Lincoln, Nebraska. "The 'Devil's Auction' deals with all people and all climes and with all history, both sacred and profane. It introduces scenes from Milton's *Paradise Lost* and scenes in a patrol wagon. The only way to define the limits of the plot is to say that it deals with the works of nature and that the scene is laid in the universe. The actors and their acts were numerous, and all the many specialties were executed and got out of the way with admirable quickness and dispatch." It was "a kaleidoscope in which the greatest number of combinations and colors are produced in the shortest possible length of time."[34] Cather admired the efficiency of the *Devil's Auction*, but it consisted of entertaining non sequiturs. A decade later, the enormous Hippodrome opened in New York with another potpourri of a performance. In "A Yankee Circus on Mars," the scene was indeed laid in the universe. It included a fat lady, a wild man from Borneo, a small town in Vermont, elephants wearing goggles and driving automobiles, a gleaming airship, and the "barbaric splendors" of Mars, including its king, aristocrats, milkmaids, and 150 dancers.[35] The performance scrambled symbols, stories, and beliefs. As the theatrical kaleidoscope turned, it rearranged familiar elements in endless variations. This description also applied to vaudeville performances, where many separate acts were combined in a program. A newspaper in Charlotte, North Carolina, observed, "People of the vaudeville stage form a human kaleidoscope. Each part or each one is a distinct unit but by various combinations they produce new effects."[36] Adding new performers to an ensemble made possible novel combinations and contrasts. The metaphor of the kaleidoscope occurred in descriptions of many sorts of performances.

Ralph Waldo Emerson argued that this constant rearrangement of familiar elements applied to the history of inventions, including the kaleidoscope itself. "The Patent-Office Commissioner knows that all machines in use have been invented and re-invented over and over; that the mariner's compass, the boat, the pendulum, glass, movable types, the kaleidoscope, the railway, the power-loom, etc., have been many times found and lost, from Egypt, China and Pompeii down; and if we have arts which Rome wanted, so also Rome had arts which we have lost; that the invention of yesterday of making wood indestructible by means of vapor of coal-oil or paraffine was suggested by the Egyptian method which has preserved its mummy-cases four thousand years."[37] From this perspective, the rotation of the kaleidoscope provided a model for history itself, which became not the story of progress but a cyclical narrative of continual discoveries, losses, and rediscoveries.

In short, between circa 1876 and 1929 there were two contradictory ways that Americans used the metaphor of the kaleidoscope. Liberals and progressives used it to describe rapid industrial change as a rich panoply of scenes that evolved quickly and were pleasing to behold but difficult to comprehend. Likewise, hopeful champions of women and those who were discriminated against, like Bowen, saw the whirl of change as the harbinger of greater opportunity. This was the most common usage. In contrast, writers such as Emerson, Twain, and Cather agreed that the surface of American culture was so engulfed by change that it seemed kaleidoscopic. However, they saw not progress but a distracting swirl of representations. Toward the end of the period, a few thinkers combined these contradictory positions, notably Henry Adams, who described technological change as a whirlwind rushing toward disaster. In his view, human striving accelerated the pace of events in the short term, but in the long term the laws of thermodynamics dictated that life on earth was subject to an unavoidable entropy and the heat death of the universe.[38] This was a minority opinion before World War I, but his autobiography became a surprising best-seller in the 1920s.

The imaginary of a "kaleidoscopic America" flourished from the centennial of American independence in 1876 until 1929, or the time it took for the great energy transition to unfold. Historians have subdivided this history into four distinct periods between Reconstruction and the Great Depression: the Gilded Age, the Progressive Era, World War I, and the 1920s. By concentrating on brief time segments, the Gilded Age is usually discussed in terms of economics, while the Progressive Era is about reform, followed by foreign affairs and World War I, and finally the "Roaring Twenties."

Chopping history into many short eras hides long-term trends. This book argues that these four periods are better understood as an energy transition between two distinct lifeworlds. It rejects this subdivision of the years from 1876 until 1930. How one divides up the past is all too often taken for granted. As Philip Scranton and Patrick Fridenson have argued, periodization tends to become a rigid framework that defines the narratives which are possible to make sense of each era. Periodization is "controlling because shifting the chunks, stitching some back together, or reslicing others proves to be terribly difficult."[39] The definition of a historical period risks becoming reified by the specialization of historians. Conferences are held and journals are created that focus on each period. Research questions focus within these short eras, and history departments hire specialists to cover each one. Moreover, "in US history, presidential terms have become templates for general histories," which leads to the idea that "elections are turning points. Yet of course they often are not." By subdividing each period into four-year presidential units, long-term trends are obscured. Major changes in work, consumption, technology, leisure, life expectancy, or identity are often not part of political programs, and they occur not in four-year segments but over decades. Legislatures commonly fail to anticipate technological and social transformations but seek to regulate new situations, new practices, or new industries after they have emerged. Politics is often reactive, and at times reactionary, and presidential politics were not the driving force in American history between 1876 and 1929.

This volume treats the years from the Centennial Exposition of 1876 until the stock market crash of 1929 as a single period when the American nation was fundamentally transformed. Presidential politics does not explain how or why this occurred. An energy transition does. Politicians did not invent (and they often did not understand) the technologies of the railroads, electrical utilities, assembly lines, automobiles, motion pictures, radio, or the new products of consumer society. Corporations, not politicians, decided where to locate the new factories, and in doing so reshaped the nation's economic geography. Local businessmen and elites, not presidents, decided when and where world's fairs would be held, and no world's fair was ever held in Washington. The work week gradually shrank by 30 percent, but this change largely came about despite presidents not because of them. Shorter hours became possible because labor organized into unions, and because work was becoming more efficient. Where the preponderance of jobs had been agricultural or blue-collar industrial in 1876, by 1929 a vast white-collar sector had emerged.

As the economic historian Robert Gordon has argued in *The Rise and Fall of American Growth*, after 1870 a period of extraordinary economic development began based on unprecedented gains in efficiency and access to new energy sources. Gross national product increased and the standard of living improved rapidly.[40] Legislation that reduced industrial working hours was based on the successes of innovative firms such as International Harvester. Laggard firms, such as the United States Steel Corporation, held on to the twelve-hour day, despite strikes demanding more modern practices. In the steel strike of 1919, the Wilson administration supported management, and the Harding administration was no better, using the courts to undermine strikes in 1922. All too often, government failed to support shorter working hours or higher wages, as it clung to an outmoded conception of the economy. This is not surprising, because until the 1920s rural America elected majorities in both state and national legislatures.

Not the government but an industrialist, Henry Ford, doubled wages to $5 a day, shortened working hours, and introduced the five-day work week. To understand why he and many other employers made such changes requires knowledge of the histories of technology, labor, and business, which are more central to understanding this half-century than politics.

Fundamental transformations in daily life such as shorter working hours, higher wages, and more leisure time did not emerge from statehouses but from the industrial economy. Progressive reformers played a role, but the changes in working hours and wages were possible because of gains in efficiency. Central developments during these years—including the consolidation of national markets, the expanding culture of consumption, the establishment of corporate research, rapid urbanization, and the invention of media events—cannot be understood by subdividing history into four-year election cycles and focusing on political debates. It is far more useful to focus on the great energy transition.

On the discursive surface of events. one finds the kaleidoscopic metaphor that many who lived during these decades used to describe the transformations they experienced. Just as a kaleidoscope keeps rearranging its contents, suggesting mutual influences and complex relationships, Americans were rearranging cultural elements to produce new patterns. The metaphor proved useful in describing rapid change. In contrast, due to the industrialization of agriculture, the political power of rural America waned, and it became fertile ground for third party movements, fundamentalist religion, immigration restriction, and racial repression. Politics in the hinterlands reacted to the dynamism of industrializing, urban America.

The kaleidoscopic metaphor was a symptom. It was less an accurate description than a hopeful image. It referred to much more than race and ethnicity, including the arts, business, entertainment, material culture, consumption, tourism, and new technologies. The metaphor was also applied to the fragmentation of time and space, the commodification of sound and image, and the reconstruction of landscapes. Yet ultimately, "kaleidoscopic" was more an obfuscation than an explanation.

2

THE GREAT ENERGY TRANSITION

Smil observes, "Clear divisions into specific energy eras are unrealistic not only because of obvious national and regional differences at the time of innovation and the widespread adoption of new fuels and prime movers but also because of the evolutionary nature of energy transitions."[1] In the United States, between 1876 and 1929 Americans witnessed but did not fully comprehend three overlapping energy transitions: (1) the completion of a shift from water mills to steam power that had begun at the end of the eighteenth century but only reached maturity after 1875; (2) the adoption of electricity to produce power, heat, and light, as well as to operate controls such as thermostats and alarms; and (3) the adoption of oil and gas to drive vehicles, heat and light buildings, and operate tools and machinery such as chainsaws, pumps, and other devices. Collectively, they constituted an unprecedented infusion of power.

The chronology of the transition was not identical for steam, oil, gas, and electricity, but each passed through the five phases that Hughes described: invention, trials, expansion, momentum, and maturity. Each of these forms of energy was developed by competing firms that merged into a few large corporations in the second and third phases of their development. In the 1870s steam power dominated transportation and surpassed waterpower in manufacturing to reach the fifth stage of maturity. Its apotheosis came at the 1876 Centennial Exposition in Philadelphia, when the great Corliss engine was ceremoniously started on the opening day. By comparison, in the 1870s the oil business had moved from the first phase of exploration and invention in the 1860s to expansion and consolidation that increasingly was controlled by Rockefeller's Standard Oil. The oil industry expanded slowly until 1899, when consumption of all forms of oil was 6.4 million barrels a year. By then it had resolved bottlenecks (or "reverse salients" in Hughes's terminology) in refining, storing, shipping (including pipelines), and distribution. Spurred by the emerging automotive industry, oil then reached its moment of technological momentum after 1900 as sales surged to ninety-one million barrels in 1909. By 1920 the oil industry had attained the fifth mature stage and was selling three hundred million barrels a year. Also, the annual miles per person traveled by automobile surpassed the railroads.[2] By 1929 oil had displaced coal as the most important fossil fuel.[3]

Electricity made the transition more quickly than either steam or oil. The electrical industries grew from a few telegraph supply firms in 1870 to a $200 million industry by 1900. The telephone, phonograph, arc light, incandescent light, dynamos, motors, and alternating current were invented and demonstrated between 1875 and the late 1880s. All of these were constantly being improved, and by 1890 the industry had moved beyond the trial installations of the second phase and began to expand beyond illumination into street railways, factories, and hydroelectric equipment. The fifteen sizable electrical firms that existed in 1885 were rapidly consolidated and merged into just two corporations, General Electric and Westinghouse. They both grew rapidly, and electrification as a whole achieved technological momentum by 1910, though momentum varied somewhat. Lighting reached that point first, street railways second, and electrified factories third. Most urban homes were only electrified after 1910. By the 1920s steam power, oil, and electrification had all reached the mature stage.

This great energy transition was far more than a change in power sources. It transformed the structure of everyday experience, leading to a new life-world. The transition concerned not only the sources, production, storage, and use of energy, but also complex changes in life's textures, sounds, smells, sights, and habits. Much of what was taken for granted in 1875 disappeared or was irrevocably altered by 1929. To take a single example, in 1875 the city street echoed with the clip-clop of horses. Its air was filled with dust and flies attracted by the horse droppings. Streets welcomed playing children, vendors with pushcarts, and a welter of pedestrians, with only minimal traffic control. The streets were weakly illuminated by gas each evening, once lamplighters lit them, one by one. When stores closed, their windows went dark and there were only a few gas-illuminated signs to brighten the scene. Few people were about. In 1929 the same street was paved, and there were no horses or manure. Pushcarts were few, and children were prohibited from playing in the street, where trucks and automobiles moved swiftly and were controlled by electric traffic lights. The street was far brighter with powerful electric streetlights, bright shop windows, and many illuminated signs. The buildings were higher than in 1875, because steel construction and elevators made taller apartment and office buildings accessible. More people were on the streets, because they had more leisure time and more money to spend, and they enjoyed a range of amusements that had not existed in 1876. Moreover, the typical city had more than doubled its population. As this example suggests, the energy transition facilitated a comprehensive transformation in street life. The changes in every other aspect of society were just as sweeping. If someone from 1875 were suddenly transported to 1929, the change in lifeworld would have as striking as a visit to another country.

Energy transitions are not sudden but incremental. The shift from waterpower to steam power or from coal to oil and gas required more than half a century.[4] This does not mean that waterpower or coal ceased to be used; they often were repurposed to generate electricity. In Smil's analysis such transitions require "decades from the beginning of production to become more than insignificant contributors, and then another two to three decades before capturing a quarter or a third of their respective markets."[5] Smil also makes clear that energy transitions were not identical from one nation to another, as each has access to diverse energy sources, faces different geographical constraints, develops divergent patterns of consumption, and creates different state and local laws. An energy transition is a slow socio-technical process that requires about half a century. It is a many-sided transformation that varies from one context to another, and it is not deterministic. Each society selects between various technical possibilities to shape systems based on its geography, history, and politics. The results are heterogeneous and contingent. Mastery of new energies ensures change but not convergence.

There are many kinds of transitions.[6] Some primarily involve production, for example, when steam power generation shifted from coal to gas, or electricity production shifted from burning fossil fuels to wind turbines. In these cases, consumers did not need to make major changes. A full energy transition involves energy production, distribution, and consumption, and it affects almost everyone in society. For example, when horses were an important source of energy, enormous investments and a large workforce were tied up in their feeding, housing, and medical care. Almost every adult had to know how to deal with horses. Production, use, and repair of saddles, riding clothes, harnesses, wagons, buggies, and horse-drawn equipment was an essential part of every local economy. The energy transition to motors destroyed those businesses and made ownership of horses and knowledge of them unnecessary for most people. That was a total energy transition, where one system replaced another, making stables, feed stores, manufacturing, and practical knowledge obsolete. Few skills were transferable to driving and maintaining an automobile. The shift to electrical heat, power, and light also had widespread consequences, whether in the home, office, factory, or public space.[7] A transition in both production and consumption develops a new material culture that affects employment, architecture, domestic arrangements, land use, and the organization of everyday life. It also suggests new conceptions of space, time, and identity.

The rapid decline of horses, and indeed the word "transition" itself, might suggest, inaccurately, that all the older forms of energy were being replaced. This was not so. Rather, the total energy available exploded in a great energy transition between 1876 and 1929. The consumption of most forms of energy increased even when being surpassed. Electric lights replaced gaslight, but consumption of natural gas more than doubled between 1919 and 1929. Sales of oil and gas combined only surpassed sales of firewood in 1916 but grew so rapidly that they surpassed coal in 1930.[8] Yet consumption of coal only stagnated briefly in the early years of the Depression and then continued to increase. The great transition was not so much from one form of energy to another, but rather a radical expansion of all forms of energy.

The multiplicity of energy forms made possible new applications of power, better wages, and a higher standard of living. Homes were linked to networks for water, sewage, heat, light, and communications. New kinds of transportation accelerated the circulation of people and goods and relocated centers of production. The flexibility and variety of energy forms also made possible the redesign of factories and the transformation of almost every job. They made possible new forms of leisure and entertainment, including phonographic music, amusement parks, movie theaters, radio broadcasts, automobile racing, and much more. The abundant new energies were used to disrupt economic hierarchies, redefine or eliminate old businesses, create new industries, and transform the means of production. Beginning in the 1890s electric streetcars carried millions of passengers, and their tracks defined the ideal locations for department stores, theaters, and offices. Just twenty years later, Americans adopted automobiles and were moving to suburbs. Electrical lighting, elevators, pumps, and telephones made skyscrapers practical. Electric motors emitted no polluting smoke, making them ideal for use in subways or inside buildings. Gasoline motors drove tractors and automobiles, transforming farming and country towns. The transition changed almost everything, and it handsomely repaid investment, creating fortunes based on oil, gas, electricity, automobiles, and mass-produced consumer goods.

Yet the great energy transition was neither planned nor directed by government, which only regulated new technologies once they were put in place. One can visualize the government's role in terms of electrical wiring. In New York City a tangle of electrical, telephone, and telegraph wires had been installed between buildings and over the streets, until by circa 1890 a vast cobweb of wires hung in the air. The rapid erection of this interwoven jumble expressed unbridled competition that continued beneath the streets as well in a tangle of sewers, water lines, gas pipes, and electric cables. In the absence of national policies, these intertwined services had to be regulated by city governments. Cities and corporations electrified with only sporadic direction and little inspiration from Washington. The national government did not take the lead. There was no federal department of energy, and the development of the energy transition was largely left to state government, private enterprise, and the cities.

THE CITY

The great energy transition concentrated industry and population in cities. In contrast, during the era of waterpower the placement of factories had been dictated by the hydrology of streams. There were some cities based on waterpower, such as Lowell, Massachusetts. But when fully developed, the Merrimack River and its tributaries drove nine hundred mills in smaller communities. Five hundred of them averaged only forty-five horsepower. Waterpower dispersed industry into the countryside wherever there was sufficient falling water. Canals also dispersed jobs and spread development along their banks. Railroads did this to a lesser degree, stimulating growth at the stations, but not all along the rails. Anyone might navigate on a canal, but railroad lines were monopolies. As Christopher Jones explored in *Routes of Power*, the ways energy is transmitted can have quite different effects. Coal shipments by water or rail stimulated local economies along their route, but neither pipelines nor electrical lines had this effect. They transmitted only one product, usually in only one direction, with nothing coming back.[9]

Local economies at each end benefited, but they established no new paths for other businesses. Pipelines and transmission lines sped urban concentration.

Water-powered factories were dispersed, but steam-powered factories could be built wherever coal could be cheaply delivered: at seaports, on navigable rivers, along rail lines, or near pipelines. Electric power made the layout of factories more flexible, and they could be built almost anywhere. The size of water-powered mill towns had been limited by available streams. Steam power released manufacturing from the geographical constraints of waterpower and concentrated workers at fewer locations. The Corliss steam engine was widely adopted because of its fuel efficiency and ability to adjust to sudden changes in load. It tipped "the balance in favor of steam, particularly in and around New England," and "fostered the growth of the big industrial urban centers," which became increasingly advantageous because of their labor pools, better transportation, and proximity to markets.[10] Water-powered cities like Lowell gradually added stream power, while cities without significant waterpower like Boston, New York, Chicago, or Atlanta built coal-fired factories. The steam hammer and steam piston used steam directly, but usually it was transmitted through gears, drive shafts, and belts. In the 1890s electric motors did away with mechanical transmission and began to revolutionize factory organization. Equally dramatic changes occurred when new energies were incorporated into mining, agriculture, communications, and business.

Table 2.1
Population growth, selected American cities, 1870–1930

City	1870	1930	Increase (%)	Density, 1930 (square miles)
Detroit	80,000	1,569,000	1,961	11,375
Minneapolis	30,000	464,000	1,546	8,382
Chicago	299,000	3,376,000	1,129	16,723
Cleveland	93,000	900,429	968	12,718
New York	942,000	6,930,000	735	23,179
Boston	251,000	781,000	311	17,795
Baltimore	267,000	805,000	302	10,227
Philadelphia	674,000	1,951,000	289	15,242
St. Louis	311,000	822,000	264	13,475

Note: Gibson, "Population of the 100 Largest Cities," tables 10 and 16.

The new urban energy systems facilitated two generations of frantic growth. Between 1870 and 1930 the population of the nation tripled, and cities increased by 700 percent, from 9.9 to 69 million.[11] More than half the nation's population was urban for the first time in 1920. The scale of growth required continual construction, as the need for roads, sewers, public lighting, housing, schools, hospitals, hotels, and offices mushroomed. The fastest urban growth between 1870 and 1930 was in the midwestern cities of Detroit, Minneapolis, Cleveland, and Chicago. Their populations grew by approximately 1,000 percent. The older cities of the eastern seaboard—Baltimore, Boston, and Philadelphia—roughly tripled in size. But perhaps the most impressive growth was in New York City, which already was the largest in 1870, yet nevertheless grew 735 percent by 1930. Every year, each of these cities needed new homes, schools, roads, businesses, and everything else for thousands of new residents. Nor was growth mere duplication. Entirely new kinds of housing, stores, factories, and infrastructure were built. Cities replaced horsecars with light rail and subways. They tore down four-story buildings and erected skyscrapers. They built new universities, museums, stadiums, railway stations, and bridges. New high-energy infrastructures penetrated everywhere. Police call boxes and fire alarms were spread throughout the city. Networks of electrical wires spread to every room in every building. Each city required an efficient network of roads, trolley tracks, subways, and rail lines to carry citizens to work and to deliver food and fuel. Their streets also contained networks of cables for the telephone, telegraph, and traffic signal systems. Cities demanded vast reservoirs, pumping stations, and systems of pipes to deliver potable water. As early as 1872, Chicago used thirty-six million gallons of water a day, and its demand increased more than 1,000 percent by 1929. Enormous fire departments were also necessary, as well as pressurized water supplied to fire hydrants on every street. Cities also needed a comprehensive sewer system, a small army of sanitation workers, and enormous dumps or landfills for mountains of daily waste. This new infrastructure was driven by steam engines and motors, interlinked with electrical switches and controls. The modern city was a network of machines driven by new energy systems that had to function continuously. This infrastructure was essential for the emerging high-energy lifeworld, yet much of it was hidden from view.

To continually expand that infrastructure, for decades the cities were vast construction sites. Many choices had to be made. Even street lighting was a complex question, with many systems on offer. In the 1880s now forgotten "moonlight towers" rose two hundred or more feet above Detroit, New Orleans, Kansas City, Denver, and many West Coast cities, only to be replaced with more numerous, shorter lighting standards a decade or two later.[12] Different arc light systems battled against gas lighting. As late as 1900, it was not obvious which to choose. Cincinnati found it necessary to send out a committee to inspect ten operating illumination systems and report on which was best.[13] Electricity triumphed only slowly in streetlighting because gas lighting became more efficient. In 1900, on most streets in most cities, arc lights were the exception and gaslight was normal. Even in fast-growing Chicago, half the public lighting in 1913 still came from gas or gasoline lamps.[14]

Not only lighting but many other improvements were made possible by the great energy transition. The new energy systems enabled new ways to live in the city. Department stores were brilliantly lighted, enormous emporiums that carried shoppers effortlessly up escalators to see vast displays of goods, to dine in comfort, and even to see art exhibits or hear lectures. Wealthier urbanities could live in luxury apartments that incorporated elevators, steam heating, telephones, electric lighting, and flush toilets. They had amenities pioneered by the best hotels, including lobbies, dining rooms, and servants. Less lavish apartment buildings introduced compact living arrangements such as beds that folded up against the wall or collapsible tables. Skyscraper offices offered executives spectacular views and concentrated a hive of white-collar workers in a single building. Such structures were only possible with reliable energy systems to supply elevator service, heat, light, water, telephones, ventilation, and power far above the street.

Improvements in infrastructure were celebrated on postcards and in guidebooks, notably the new subway systems built in Boston, New York, and Philadelphia; the elevated lines in Chicago; and new lighting systems, parks, and waterworks. More than seventy-five civic improvement associations developed into the City Beautiful movement, which worked to improve public spaces that all classes could enjoy on an equal footing, such as public libraries, museums, and swimming halls. They considered streets to be more than transportation arteries; they wanted boulevards wide enough to include trees and outdoor cafes where citizens commingled. They believed intensified lighting combated crime, reduced accidents, and drew residents into public space at night. They lobbied for ornamental light standards and "white ways." They viewed the great expositions in Chicago, St. Louis, and San Francisco as models for an achievable urban future, and they supported plans that could improve social cohesion, order, and civility. In that era magnificent public libraries were erected, the YMCA and YWCA flourished, and many museums were established. The public believed that enhancing the urban environment improved public health and safety, created shared spaces for community, and increased civic pride.[15]

For transportation, the city of 1876 had relied almost exclusively on horses, which were replaced by streetcars, trucks, and automobiles. The city of 1876 burned gas for lighting and coal for heating, fouling the air; the city of 1929 was lighted by electricity and burned coal more efficiently. The streets were asphalt or concrete, and traffic was automatically regulated. The largest cities built subway systems beneath the business district. And where commerce in 1876 had been dispersed into hundreds of small shops, in 1929 it was concentrated in department stores and large specialized stores.

One obvious measure of the change was in the height of buildings. In 1876, the highest towers in American cities still belonged to churches, notably Trinity Church at 79 Broadway in New York, the highest building in the city from its erection in 1846 until 1890. At 279 feet, it was the tallest building in the United States until surpassed by Saint Michael's Church in Chicago. Both were dwarfed in the 1890s by skyscrapers owned by newspapers, insurance companies, the Singer Sewing machine company, the Wrigley's chewing gum corporation, and, in 1913, the Woolworth Building, which at 792 feet was the world's tallest structure. Popularly called "the cathedral of commerce," it was three times the height of Trinity Church. Reformers successfully prevented New York skyscrapers from towering straight up from the sidewalk, and setbacks were required to allow more sunlight to reach the streets. Nevertheless, taller buildings were erected, culminating in the 1,250-foot Empire State Building, completed in 1931. Skyscrapers formed the dramatic, secular skyline that became the hallmark of American cities. This skyline visualized the great energy transition. Movement accelerated; lighting flashed ever brighter; buildings soared ever higher. The city literally roared, in a constant barrage of noise. It naturalized new experiences, phenomenal crowds, alluring commerce, a huge variety of entertainments, and new kinds of work.

WORK

The new sources of energy were used to change almost every kind of work, whether on the farm, in the small town, or in the city. Although there was an overall surge in population, some towns lost residents, and some jobs disappeared. As agriculture mechanized, farmers declined in numbers and political influence. Farmers became more efficient, but their incomes did not keep pace with urban incomes. Agricultural prices peaked during World War I but fell in the 1920s. Wheat that sold for $2.19 a bushel in 1919 went for $1.05 in 1929. Cotton fell from thirty-five cents to just seventeen cents a pound.[16] Many farmers sold out. Nor were the farmers alone in decline. As the Census Bureau summarized: "In 1870, over 75 percent of the nation's labor force was engaged in the production of physical goods, that is, in agriculture, forestry and fishing, extraction of minerals, and manufacturing and mechanical industries."[17] After 1870 the trend was "mainly away from production of physical goods and toward distribution and service, toward pursuits in transportation, communication, and trade; service pursuits; and clerical pursuits." By 1930, only a little more than half the workforce produced physical goods, and many of these were in the new automotive industry. There was enormous growth in skilled trades. In 1880 there were only 1,200 electricians; in the 1930 there were 280,000. The telephone and telegraph industries in 1880 employed 57,000; in 1930, 220,000. Demand for skilled workers increased by 400 percent or more in the building trades, including painters, glaziers, plumbers, electricians, and roofers, with carpenters increasing almost as fast. These were well-paid jobs. Factory labor increased just as rapidly in automobile plants, petroleum refineries, and electric utilities. Some of this labor was repetitive assembly line work, but demand also quadrupled for skilled boilermakers, machinists, and tool makers. Many new jobs were created by the rapid adoption of automobiles, including demand for mechanics, tire factory workers, oil drillers, road builders, and other skilled tradespeople.

Growth was not universal. Some jobs were mechanized, and the number of sawyers, coopers, and glassblowers fell by more than half. Fewer people worked in quarries, brickyards, or cigar and cigarette factories. Increasingly, machines such as electric cranes or conveyor belts did the heavy lifting and carrying, whether unloading ships, hauling coal out of mines, or moving materials inside factories. Inventors also created new industrial systems that transformed work. For example, Thomas Edison developed a long kiln for cement production that could produce 1,000 barrels a day. It did so while eliminating many unskilled workers. His system was so superior that "by 1910 half of the cement produced in America came from Edison type kilns, and all the larger producers started taking licenses under Edison's kiln patents."[18] Fewer workers were needed to produce a barrel of cement, and its cost fell by more than half. Something similar occurred in the steel industry. Starting in 1922, steel mills introduced the continuous strip production process, which eliminated many jobs. Before then workers used hand tongs to maneuver red-hot metal slabs through rollers and finishing machines, but this metal handling was eliminated by the continuous strip process, which moved slabs as fast as 1,000 feet per minute.[19] Thousands of jobs disappeared, but, as with cement production, other positions were created because steel became less expensive, which stimulated demand.

These changes in cement and steel production were part of a widespread substitution of machinery for human labor. It had begun with the growth of waterpower before the Civil War, which in turn was overtaken by steam power. By 1880, American factories used water wheels and steam engines to produce 3.66 million horsepower. Falling water still drove mills in Lowell, Cohoes, Rochester, and Minneapolis, but steam power was increasing. In New England, for example, steam provided just a quarter of industrial power in 1870, but more than half by 1890 and two-thirds by 1900.[20] Even so, New England's waterpower doubled during these thirty years. To become dominant, steam power increased by 1,200 percent.

The steam-powered economy had significant social costs, notably, air pollution, mine accidents, and boiler explosions, but Americans emphasized the advantages it conferred.[21] At the Centennial Exposition held in Philadelphia in 1876, the central exhibit was the world's largest steam engine. Standing forty feet tall, the Corliss engine drove eight line-shafts that powered every machine in the fourteen-acre Machinery Hall. This display presented the supremacy of steam as the culmination of the nation's history. Steam drove steamboats and railroads faster than canal boats. Steam heated buildings; it pumped water supplies. It drove hotel elevators, blast furnaces, pile drivers, enormous forging hammers, steam shovels, and all the manufacturing systems exhibited in Machinery Hall. Steam relocated manufacturing, stimulated urban concentration, linked the nation into a single market, and provided conditions that favored large corporations over small businesses and department stores over neighborhood shops. Steam made possible much of what was celebrated as kaleidoscopic America, and it remained the dominant power source until after 1900.

In these years the railroad became the basis for what cultural geographer John Stilgoe termed the "metropolitan corridor." Its nodal points lay in "the industrial zone, of the manufacturing fringe of every great American city east of the Rocky Mountains." It was "an awesome landscape of built forms" that railway passengers glimpsed on the outskirts of each city. Emerging rapidly after 1880, this new industrial landscape was the physical expression of the new energy regime. These "giant manufacturing operations required entire complexes of cavernous buildings: steel mill complexes stretched for miles along rivers, their buildings and furnaces linked by conveyors, railroads, and catwalks." The new factories were one-story, sprawling complexes that proved to be "vastly more productive than multistory, elevator-serviced structures."[22] The railroads delivered to these factories coal, limestone, wood, stone, and other raw materials and took away both waste and finished products. The metropolitan corridor included sidings packed with boxcars, cranes, furnaces, smokestacks, and mountains of ashes (figure 2.1).

As Stilgoe emphasizes, this landscape was less often criticized than celebrated, and not only in trade journals such as *Industry*. Photographers such as Alfred Stieglitz and artists such as Reginald Marsh also depicted these sites. To some visitors, a "luminosity suffused the industrial zone, one that changed intensity at twilight and other times. It enthralled discriminating observers before and after the perfection of the electric light," including Upton Sinclair and Theodore Dreiser.[23] Economists, notably Paul Krugman, call this railway corridor a "manufacturing belt" that exemplifies how "there is always a tension between the forces of agglomeration," in this case steam power and the railroads, "which tend to make activity clump together, and the centrifugal forces that tend to make it spread out."[24] Between 1870 and circa 1920 the forces of agglomeration were stronger, and they were initially reinforced by the advent of electricity.

For even before steam power reached its apotheosis, the great energy transition intensified with the adoption of electric light, heat, and power. At first, electricity was almost exclusively in city centers and therefore increased the concentration of industry there. Over the long term, however, as electrical transmission spread, it would become a centrifugal force. In the 1880s electricity was still a negligible source of factory power, and

Figure 2.1
Louis Lozowick, *Steel Valley*, 1936, lithograph. Courtesy of the President and Fellows of Harvard College.

it was only beginning to be used for lighting.[25] Once better motors were developed, they began to be used in mills and small businesses such as print shops. Companies that did not generate electricity could buy it from a utility. Both water and steam transmitted power mechanically. In contrast, because wires could easily be moved, an electrified factory could experiment with different arrangements of machines to optimize productivity. As electrification took hold, the growth of inanimate power in factories was enormous, tripling to 10.3 million horsepower by 1900 and reaching 19.5 million horsepower in 1930. Some of that new factory power came from steam engines, but most came from electric motors. Industry was by far the largest customer for utilities and equipment manufacturers such as General Electric. In 1922, factories consumed fifteen times more electricity than the residential market and nine times more than the commercial market.[26] Electric machinery could be used to replace workers and to redefine those who remained. One dramatic example was in mines, which long relied on men, mules, and horses to move coal and ore. Only 0.7 million horsepower were mechanical in 1880, but this rose 800 percent to 5.6 million horsepower in 1930. The number of miners increased until 1910 but then stagnated. Yet total coal production increased, as machines were substituted for men. By 1930 electrical wiring had penetrated all sectors of urban life, including schools, hospitals, theaters, restaurants, and other public spaces. After 1919 urban homes were rapidly wired, accompanied by a vigorous advertising campaign from General Electric, Westinghouse, and local utilities.[27]

In contrast to this electrified urban lifeworld, by 1930 electricity had only begun to reach the countryside. The further it spread, the more it became a force not of agglomeration but of dispersion. Suburbs became more attractive once electrified. Factory sites away from the center of cities were abundant and inexpensive. As Krugman explains, the combined effect of electrification and gasoline-driven cars and trucks was that "centrifugal forces gained the upper hand. Trucking made it less important to be near a rail hub, while electrification led to a redesign of factories—hulking mills replaced with sprawling one-story structures."[28] A major long-term effect of the multiple energy transitions was the de-concentration of manufacturing, which spread far more widely throughout American society.

The assembly line exemplified these trends. At its inception, it concentrated workers in enormous factories, speeded production, and made possible both higher wages and lower prices. Gradually introduced in 1913 at the Ford Motor Company in the fast-growing city of Detroit, the assembly line was the culmination of several trends (figure 2.2). First, the accuracy of machine tools had gradually improved during the nineteenth century. Eli Whitney had early championed the idea of interchangeable parts, but he promoted the goal rather than achieving it in practice. Merritt Roe Smith has demonstrated how slow and difficult interchangeability proved to be for the arms industry.[29] Success came gradually and depended on long-term financial support from the federal government. David Hounshell, in *From the American System to Mass Production*, traced how the practices of the arms industry made their way into the factories that produced clocks, harvesters, sewing machines, bicycles, and other devices that were easier to make when metal parts were interchangeable. Manufacturing became more efficient, repairs to broken machines were easier, and an assembly line became possible.[30]

Interchangeable parts were necessary, but they were not sufficient to make an assembly line. Just as important, electrification released factories from driveshafts, belts, and gears. They were rigid and lost so much energy in

Figure 2.2
Assembly line at Ford Motor Company, Highland Plant, 1913, photograph. Courtesy of the Library of Congress, Prints and Photographs Division.

transmission that all the machinery was placed as close to the power source as possible. Factories therefore had three to five stories, with the power source in the center. In an electrified factory, machines could be placed anywhere and in any order. The layout of production could easily be rearranged, because wires were far easier to reposition than gears and driveshafts. It was no accident that Ford's managers hit upon the assembly line as they moved production to a new, electrified factory. As managers reorganized production, they realized that it was most efficient to give each worker a narrowly defined task that could be completed in one or two minutes. Before the assembly line, small teams of workers had built up each car at a fixed location, and they moved around the factory looking for whatever parts they needed. In contrast, the assembly line brought both parts and partially assembled cars to the workers, who remained in one place. The work arrived at a comfortable height. There was good lighting and often a stool to sit on. A worker no longer needed to master many skills. A short, repetitive job could be learned in a short time, and one soon became efficient. Some tasks could be done by people with disabilities, including even the blind, and Ford made it a point to hire them for certain jobs. In 1909, the assembly of a Model T had taken 720 minutes; by 1914 it took only 93 minutes.[31] As this example suggests, the adoption of electricity led to radical changes in employment.[32] At first, it enabled massive centralization. But some of the largest firms, including Ford, later began to decentralize assembly, with branch plants in many locations. In 1914, all Fords were made in Detroit, but the centrifugal forces of widespread electrification, better highways, and trucking made it economical to ship parts and assemble cars in regional factories.

As the development of the assembly line also suggests, the multiple energy transitions improved efficiency but reduced demand for workers in some areas. Until circa 1910 employment in many mines and industries roughly kept pace with population growth, but then their numbers remained unchanged for decades while production kept increasing. Such gains in efficiency were notable in furniture factories, cotton mills, woolen mills, and the railroads, and to a considerable degree they were facilitated by electrification. Electrical machines were regular in their action, creating more uniform products, and better lighting enabled many enterprises to function longer in winter, as late sunrises and early sunsets no longer mattered. Annual productivity gains in manufacturing of 1.6 percent may seem small, but when compounded over fifty years from 1869 until 1919, the increase amounted to 120 percent. Then came a spectacular increase during the 1920s, when new factories, relying on electrical equipment and assembly lines, proved enormously efficient. Between 1919 and 1929 alone, productivity increased an astonishing 60 percent.[33]

Contrary to the popular saying, however, the tide of efficiency and prosperity did not "lift all boats." While the population tripled, the number of workers in some fields stagnated or declined. The number of boatmen, canal men, and lock keepers fell, as did demand for workers in agricultural implement plants, sawmills, the brass industry, beverage production, or broom and brush factories. Mechanization replaced skilled work in many areas. In 1930 there were fewer millers, hat makers, tailors, milliners, and dressmakers than there had been in 1900. More than half the harness and saddle makers, stable hands and hostelers, draymen, teamsters, and carriage drivers disappeared. Some of them became chauffeurs, truck drivers, or tractor operators.

Meanwhile, a prodigious increase in white-collar employment drew women into the workforce as clerks, typists, stenographers, and salespersons. Most notable of all was the quadrupling of creative jobs in theaters, advertising, architecture, art, music, teaching, journalism, photography, and the radio and motion picture business. There was also considerable demand for clergymen, lawyers, judges, and the police. Growth was a slower but still significant for bakers, bankers, brokers, doctors, firefighters, and librarians.

This dynamic but turbulent labor market can be understood as part of the process of "creative destruction" in the US economy. The phrase "creative destruction" was coined by Joseph Schumpeter to describe how competitive capitalism continually innovates new products and processes. He placed innovation rather than the "labor theory of value" at the center of the economy. If Karl Marx said that under capitalism "all that is solid melts into air," Schumpeter celebrated "creative destruction." He declared, "Without innovations, no entrepreneurs, without entrepreneurial achievement, no capitalist returns, and no capitalist propulsion. The atmosphere of industrial revolutions—of 'progress'—is the only one in which capitalism can survive."[34] Schumpeter praised American business executives as wealth creators and as the enemies of tradition. In contrast, he criticized the hindrances to markets in Central Europe, which grew more slowly and lacked American dynamism. He argued in favor of the economies of scale that large corporations had compared to private firms. He also concluded that monopolies were not as dangerous as many supposed, for they remained vulnerable to innovative competitors. Schumpeter concluded that entrepreneurs drove progress, undermined the class structure, and broke traditions. His entrepreneurs did not accept the idea of a steady-state economy. Rather, they looked for innovations and believed in "the insatiability of wants" from consumers.

Some companies were destroyed because rivals had new patents, such as Edison's cement kiln, or adopted new forms of organization, such as Ford's assembly line. Other companies were absorbed in a process of consolidation. There was a "high rate of business formation during the great merger wave in American business (1897–1904)." Many companies began to manufacture automobiles between 1900 and 1910, but by 1930 most had either merged or gone bankrupt. The initial florescence of small firms was followed by consolidation. By the mid-1920s Ford and General Motors controlled the lion's share of the market. Yet even with just two large competing firms, innovation continued. Similarly, a great many electrical equipment manufacturers were merged into either Westinghouse or General Electric. One might expect that oligopoly would lead to stagnation, but the electrical industries also proved particularly dynamic. Such "firms with strong market positions become powerful engines of technological progress." In both the automotive and electrical industries, small firms flourished by making parts, accessories, or appliances. In general, when new markets developed, many subsidiary inventions followed.[35]

By 1900, the process of invention was being institutionalized. Westinghouse and General Electric were among the first companies to establish full-time research laboratories. Other large American corporations also saw the advantage of funding innovation, and research laboratories emerged dealing with pharmaceuticals, communications, and chemistry. Corporations understood that patents, new materials, and new practices would enhance their market positions. In addition, independent firms such as Arthur D. Little specialized in innovation. As invention became a full-time occupation, the annual number of patents increased from 13,000 in the 1870s to 43,000 in the late 1920s.[36] New industries grew up to produce electrical appliances, radios, films, airplanes, nitrogen fertilizer, cosmetics, drugs, and much more. Not every patent proved profitable, but then again not every valuable innovation was patented. For example, Thomas Edison did not patent his invention of an X-ray machine that could help doctors see internal injuries, and Henry Ford did not patent the groundbreaking innovation of the assembly line.

As productivity increased, wages gradually rose, in part because unionized workers pressured management. Robert Thurston, president of the American Society of Mechanical Engineers, noted in 1895 that wages in textile mills and other factories had risen 20 percent in the previous twenty years, even though working hours had fallen. So rapid was the improvement, Thurston argued in the *North American Review*, that "we are fifty percent more comfortable than in 1880."[37] He correctly anticipated that these trends would continue into the twentieth century. Between 1909 and 1940, steel workers raised the value of their output per hour by 295 percent, chemical workers by more than 300 percent, and manufacturing workers as a whole by more than 250 percent.[38] Thurston, like most managers, believed industrialization was making life better for all. But how did the working classes experience these changes?

For workers, the shifts from animate to inanimate power and from workshops to ever larger factories had paradoxical effects. Machines replaced many jobs and reduced the demand for some forms of skilled labor. At the same time, wages rose and working hours fell. But work was insecure. Many industries had seasonal layoffs, and if the economy suffered a depression, workers were fired or put on reduced time. Business activity dropped by one-third or more in 1873–1879, 1882–1885, 1893–1897, 1910–1911, and 1920–1921. Work could also be halted for other reasons. A water-driven mill might close because of a drought. A strike by coal miners might create a fuel shortage and force steam-driven factories to close a thousand miles away. As the labor historian Daniel T. Rodgers summarized, the industrial worker was caught in "a boom-and-bust cycle that afflicted virtually every industry," and it almost seemed that they either worked overtime or not at all. "Such swings were most violent in the clothing trade where the year was divided into two intensely busy seasons, interrupted by twelve to eighteen weeks of slack work or none."[39] Similar ups and downs afflicted shoemakers, meat packers, cannery workers, and foundry men. Likewise, steel plants ran "at full production or not at all," and such swings in production had knock-on effects for coal miners and railroad workers. Even in the prosperous year of 1892, workers in Pennsylvania experienced at least a month of shutdowns in steel mills, coal mines, carpet mills, and window glass works.

Unskilled workers suffered most, because someone else was always ready to compete for their low wages. Skilled workers generally prospered, but they could be fired when a merger consolidated production and closed factories. As the size of corporations increased, workers became more anonymous to management, and they had to look out for themselves. Rather than stand alone, many joined unions, which affirmed their identity, valued their skills, and represented them in negotiations. Most skilled trades were organized under the umbrella of the American Federation of Labor.

In the turbulent economy, workers became accustomed to changing jobs frequently. In 1912 the Department of Labor surveyed meat-packing plants, machine shops, textile mills, and automobile factories and found that annual worker turnover averaged 115 percent. A factory with 500 workers needed to hire and train 575 people every year just to stay the same size, assuming everyone came to work. But absenteeism also was a problem. At one Ford plant, there were 13,623 jobs, but every day one in ten workers did not show up, and the company therefore needed 1,300 substitute workers.[40] Absenteeism reflected dissatisfaction with the repetitive labor on the assembly line. To reduce turnover, stabilize production, and save money on hiring and training, Ford doubled wages to $5 a day.

Many industries that adopted the assembly line more than doubled productivity. There were other innovations as well. For example, the Bessemer system improved steel manufacturing, and time and motion studies developed by Frederick Winslow Taylor imposed rational routines in some factories. Also important were incremental innovations in inventory and marketing. These changes deskilled or routinized semi-skilled work. Yet a large part of the industrial economy was not amendable to such innovations. As Philip Scranton demonstrated in a groundbreaking work, specialty production companies that made complex goods, such as elevators, turbines, switchboards, and precision instruments, could not profitably adopt mass production. They reaped no advantage from deskilling labor or from subjecting it to routines. Instead, they prized skilled workers who could respond flexibly to demands for varied goods. This was also true for companies producing high-quality consumer goods, such as carpets, furniture, jewelry, cutlery, hats, and clothing. Skilled workers made possible the endless novelty that was the hallmark of consumer society. Specialty production companies explored systems that "could increase reliability, reduce errors, speed batch throughput." But if work restructuring "went so far as to freeze flexibility and innovation," it then became "a burden and a constraint."[41] Taylor's prescriptions or Ford's assembly line were useful for the repetitive production of identical things, but they were of little use in the batch production of specialty items or luxuries. Consumers wanted variety, and large profits accrued to the firms that could supply it. Crucial to the long-term success of these firms was their ability to hold on to their workers. Specialty goods demanded skilled labor and flexible production. Accordingly, wages rose as companies competed for the best workers, even as efficiency improved and pressure from unions increased.

The factors that ensured one man's success often meant disaster for another. If an automobile manufacturer like Ford became a billionaire, those whose livelihood was inseparable from horses saw their business wither. Many who held jobs in the gas industry, particularly those who installed, maintained, and lighted gas lamps, lost their positions as electric lights were adopted. In factories, glassblowers were replaced with machines that mass-produced identical bottles. In harbors, stevedores were made redundant by powerful cranes that could lift enormous loads and swing them from the ship's hold to the dock. In mines, squat electric locomotives, elevators, and conveyor belts carried ore to the surface, replacing horse-drawn skips and their drivers. In cinemas, musicians who accompanied silent films were out of work when "the talkies" arrived with their own soundtrack. Telephone switchboard operators lost their jobs as direct dialing was introduced for local calls. Technological change eliminated jobs, leaving behind many with expertise that no one needed. The number of jobs did not decline, as some feared, but many saw their training become obsolete.

Even if one chose an occupation in a growing segment of the economy, accidents could maim a worker. The courts recognized the doctrine of "assumption of risk," which maintained that one could not expect to receive compensation for accidents that were foreseeable in a particular line of work.[42] Working with blast furnaces was a dangerous trade, and many a worker suffered "foreseeable" injury or death. Labor history is filled with stories of railroad workers losing an arm or leg, or of miners killed or maimed by explosions or collapsing tunnels. Workers might be suffocated in a grain silo, be electrocuted by touching a live cable, or lose a hand in an industrial machine. Individual accidents received at best a line or two in the newspaper. But spectacular disasters became national sensations, notably the 1911 Triangle Shirtwaist Factory fire, where 146 men, women, and children perished because the fire doors were locked. Most were Italian or Jewish immigrants. Laws passed afterward established safety standards and inspection agencies to enforce them.

There were "foreseeable accidents" in coal mines, as well as the long-term dangers of black lung disease. As with the Triangle Fire, safety legislation often came only after a particularly gruesome disaster. The 1869 Avondale Fire in a Pennsylvania mine killed 109 men who had no chance of escape because the blaze broke out near the only exit.[43] The following year laws mandated better conditions, but there was worse to come. Between 1892 and 1912 there were twelve more mine disasters in which more than one hundred men and boys died. Finally, a federal Bureau of Mines was created in 1910, which established and enforced safety standards.[44]

Awareness of the dangers of the workplace and respect for the hard work of ordinary people led to the creation of Labor Day. First celebrated in

1882 in New York, it quickly spread to other parts of the nation. It became a federal holiday in 1894 after a major strike. In May of that year, the socialist leader Eugene Debs had called a strike in Chicago at the hub of the national rail network. The dispute concerned workers at the Pullman Sleeping Car Company, whose cars were on trains throughout the country. In the economic depression of 1893, the company had eliminated 40 percent of the jobs and slashed wages by 25 percent. In response, the workers struck, and out of sympathy railway unions refused to work on trains if Pullman sleeping cars were attached. When the railroads refused to remove Pullman cars, the nation's rail system ground to a halt. The attorney general sent in the army to break up the strike. But it persisted long enough to show the power of collective action. The adoption of Labor Day that year was a symbolic victory for labor, signaling the growing importance and usefulness of unions.

The rise of unions constituted a rejection of individualism as a misleading and divisive ideology. Unions responded to the concentration of capital in corporations by concentrating the power of labor. A fragmented labor force of self-reliant loners could not bargain for fair wages and good working conditions. Even workers in dynamic sectors of the economy could only receive wages that kept pace with rising productivity if they stood together in solidarity. A united front against management ensured that negotiations would be taken seriously and that strikes could be won. Unions also supported reformers who successfully campaigned for abolition of child labor and against dangerous working conditions. Yet in the United States unionization was not accompanied by development of a large labor party. Why not?

When the German sociologist Werner Sombart visited the United States, he sought to understand why American workers were not organized into a political force as a labor party. In 1906, he published *Why Is There No Socialism in the United States?* He adduced three reasons. First, the long established two-party political system blocked socialist parties from power because they seldom could muster more than half the votes to win an election. Sombart reviewed the history of third parties and, noting the fate of the Populist Party in the 1890s, realized that if an upstart party was at all successful, some aspects of its program were adopted by one of the two major parties. The third party then faded away. Labor historian David Montgomery later concluded that socialists in the United States focused on control of the workplace and on municipal government, rather than on establishing national political power. Workers were highly aware of local political machines and their corruption, patronage, and power, and at times the workers' candidates gained control of a municipality, for example, Schenectady or Milwaukee.[45] Contending with local conditions pushed them toward pragmatism.

The second, and the main reason that socialism did not fare well in the United States, Sombart argued, was because the American workers were paid higher wages. Immigrants from Germany had a better material life than their cousins back home. They dressed better, they ate meat more often, and more of them could afford to buy homes and furnish them. Workers saw these material improvements as proof that their lives were getting better. Moreover, they had greater freedom of speech, and their children could rise to a higher-class position if they did well in the free public schools.

In a short, final section of his argument, Sombart noted that some workers believed that they could leave industrial work and go west to acquire a homestead. Yet this migration was more feasible before 1880. By 1900 the best lands for homesteading had been taken up, and much of what was left lay in arid zones where costly irrigation would be necessary.[46] However, Sombart realized that Americans moved frequently, and the idea of homesteading continued to have powerful appeal. Moreover, there was an overall shift from East to West. In 1880, only 1.8 million white people lived in the West; by 1930 there were 11.8 million.[47] Some of these migrants succeeded in developing irrigated farms, but most settled in cities and towns.

Other factors must be included for a more complete explanation of the weakness of socialism in the United States. One is the animosities immigrant workers brought with them from Europe. There was no love lost between the English and the Irish, or between the French and the Germans, or the Italians and the Austrians. Likewise, labor organizers confronted the historical animosities within the Balkans and between Eastern Europeans. Even when immigrants did not come from antagonistic nations, they spoke different languages, practiced different religions, and lived in separate neighborhoods. Eric Foner has suggested that the question of socialism in America is a distraction from a more intriguing possibility: "Perhaps, because mass politics, mass culture, and mass consumption came to America before it did to Europe, American socialists were the first to face the dilemma of how to define socialist politics in a capitalist democracy."[48]

If the Socialist Party remained small, however, hundreds of utopian and dystopian books and articles appeared, each convinced that the nation was on the cusp of fundamental change. Their popularity suggests that many readers sensed their lifeworld was changing. In 1888, Edward Bellamy's *Looking Backward* was one of the most influential.[49] It described a vastly improved economic system of the future, where wasteful competition had been eliminated, mass production lowered prices, and all citizens could purchase whatever they needed from enormous department stores. Poverty had disappeared amid a general equality. Bellamy's book sold more than 400,000 copies and inspired many imitators. It also led to the formation of clubs dedicated to the realization of the society he described.[50]

Another utopia was projected by King Camp Gillette, who had made a fortune by improving and mass producing the safety razor that still bears his name. Gillette's *The Human Drift* argued that the United States had entered "the most critical period in the history of this country" when a splendid new society might emerge "in place of our present system of competition," a society based on what he called "material equality." For the rest of his life, he would "advocate the formation of a united stock company by the people, of sufficient magnitude to gradually absorb and finally control production and distribution, such company having in view the destruction of all tributary industries of the present system."[51] He imagined creation of an enormous city, Metropolis, sixty miles long and thirty miles wide along the shores of Lake Erie and Lake Ontario (figure 2.3). The waterpower of Niagara Falls would drive the machinery of a single, efficient corporation that had absorbed all others. All transportation would be by electric vehicles with rubber tires. Many other utopian works between 1870 and 1930 based their ideal societies on improved technologies of production and new sources of energy. In Gillette's Metropolis sixty million "people would live in mammoth apartment houses or hotels and be free from all the annoyances of housekeeping. These apartment buildings would be conducted upon a scale of magnificence such as no civilization has ever known."[52] Each of the 36,000 buildings was to be of steel beam construction modeled on the new office buildings in Chicago and New York. They would be similar but not identical. They were to be twenty-five floors high and circular in form, with dining areas in the center, but otherwise distinctive, using "all the prismatic tints of the rainbow" and "kaleidoscopic effects in colors and designs."[53] Gillette was but one of many enthusiasts for using modern materials and mass production to raise the level of everyday life. William Dean Howells explored similar ideas in 1894, in a novel that contrasted the

Figure 2.3
Apartment buildings for proposed Utopian City near Niagara Falls. King Champ Gillette, *The Human Drift* (Boston: New Era Publishing, 1894), plate 4, p. 102.

United States with an imaginary island nation, Altruria, that had achieved economic and social equality.[54]

In contrast to such visions of future harmony, dystopian texts extrapolated from class divisions and tensions. For example, in 1905 Ambrose Bierce published a bitterly satirical work supposedly written several thousand years in the future, when the United States had ceased to exist, having descended into anarchy and corruption. "Beginning with the slaughter at St. Louis on a night in the year 1920, when no fewer than twenty-two thousand citizens were slain in the streets and half the city destroyed, massacre followed massacre with frightful rapidity. New York fell in the month following, many thousands of its inhabitants escaping fire and sword only to be driven into the bay and drowned. . . . A week later Washington was destroyed, with all its public buildings and archives; the President and his Ministry were slain, Congress was dispersed, and an unknown number of officials and private citizens perished."[55] Bierce described democracy as a flawed form of government doomed to mediocrity. In *The Iron Heel* (1908) Jack London also imagined a future where class divisions were severe, and the socialists grew powerful. That so frightened the wealthy oligarchy that they seized power, eliminated democracy, reduced workers to slavery, and ruled through terror. London imagined the establishment of a brotherhood of man would only come centuries later. Neither the utopian predictions of abundance nor the dystopian fears of chaos proved accurate, but these contrasting narratives underscored the uncertainties aroused by the changing lifeworld during the great energy transition.

MOBILITY

If new energy forms made plausible visions of radically different futures, more immediately they enabled rapid urbanization and the gradual depopulation of the countryside. If energy transitions facilitated the transformation of the hours of work and the jobs most people performed, these transitions had an even greater influence on how they traveled. Americans had long been avid travelers, moving about the continent to a degree unknown in Europe. Tocqueville noted this tendency already in the 1830s. "In the United States, a man builds a house in which to spend his old age, and he sells it before the roof is on," and throughout life the American was in "a bootless chase of that complete felicity which forever escapes him."[56] James Bryce concurred in 1910, declaring, "They are an unsettled people," continuing, "nobody feels rooted to the soil. Here today and gone tomorrow."[57] This restlessness was abetted by the adoption of steam-powered railroads, electrical trolleys and interurban lines, bicycles, and the automobile. Sombart's explanation of why socialism did not flourish in America rested partly on his realization that Americans did not like to stay put.

As they adopted new modes of transportation, Americans experienced their society as an unfolding spectacle. They had begun to do so when riding the railroads, with plate glass windows framing the passing scenery. They next adopted bicycles after 1877, streetcars after circa 1890, and automobiles after 1900. Each of these technologies offered a different travel experience. Each widened the distances one could travel with ease, and each changed the relationship between the individual and the community.

The bicycle emerged as an American fad in the late 1870s after British cycles attracted attention at the Philadelphia Centennial Exposition. They did not resemble the modern article but were high wheelers. The pedals were directly attached to a high front wheel that was four feet or more in diameter. This metal wheel was rimmed with solid rubber, and the cyclist sat above it, while a much smaller wheel trailed behind to improve stability and balance. Each cycle was hand-made, with the wheel's size determined by the length of the purchaser's leg. There were no gears. High wheelers were expensive, costing as much as half a year's wages, and a common laborer could not afford one. In 1878 a Columbia "ordinary cycle" cost $313. In 1884, Thomas Stevens, an Englishman, rode a Columbia "Expert" from San Francisco to New York City, becoming the first person to cross the continent on a bicycle. He then rode across Europe, the Middle East, and Asia. Stevens traveled light, with only spare socks, an extra shirt, and a slicker that doubled as a tent. The early high wheelers were ridden by middle- and upper-class men and a few women, sometimes forming parades of cycles, for example, in Hartford in 1885 (figure 2.4). In 1888 the Philadelphia Tricyclists Club's members included one hundred men and eighteen women, one of whom won the prize for riding the furthest during the year: 3,304 miles. These enthusiasts were not confined to city streets but soon laid out "wheelways" connecting city parks, traversing the countryside, and stimulating touring by bicycle. New York's "Ocean Parkway" that linked Brooklyn's Prospect Park with Coney Island was opened to cyclists in 1895. Other dedicated bicycle paths were laid out through parks and along waterways in Boston, Buffalo, Cleveland, Minneapolis, Philadelphia, and other cities.[58]

Women seeking greater freedom of movement embraced the bicycle. At first, some adopted the high wheel tricycle, which had pedals attached to a chain that drove two large back wheels. Many more joined in once the modern bicycle had been introduced in the 1890s, with two wheels of equal

Figure 2.4
Bicycle parade on Main Street, Hartford, Connecticut, 1885. Frank Hale, photographer. Courtesy of the Library of Congress, Prints and Photographs Division.

size, a chain drive to the back wheel, and inflated tires that gave a less jarring ride. The bicycle craze helped to abolish the use of the bustle and corset. It encouraged "common sense dressing" that released women from physical constraints on their movement. Some conservatives thought women cyclists indecently exposed their legs by wearing shorter skirts. Some physicians declared that riding a bicycle stimulated females sexually, promoted immodesty, and harmed their reproductive organs. But the suffrage leader Susan B. Anthony declared in 1896: "Bicycling . . . has done more to emancipate women than anything else in the world. I stand and rejoice every time I see a woman ride by on a wheel. It gives woman a feeling of freedom and self-reliance."[59] In the 1890s manufacturers created safety bicycles that were easier for women to mount, protected their clothing from being caught in the drive chain, and absorbed shocks with springs under the seat.

By 1900 five million bicycles were in use,[60] and their cost had fallen so much that they no longer marked class differences. A mass-produced safety bike could be purchased for a month's wages or less, and a market had emerged for used bicycles. Cyclists organized clubs and associations that pushed for better roads well before automobiles became common. The bicycle increased the range of movement, and it was easier and less expensive to own, maintain, and ride than a horse. Where one previously walked within the radius of a mile or two from home, one could now roam with ease at least three to five miles in every direction. Moreover, bicycles could carry shopping in baskets, making it easier to transport purchases. Messengers on bicycles became common in cities, and in World War I troops rode specially designed bicycles that could carry heavier loads. Anyone born in 1870 understood that the bicycle had transformed mobility for millions of people and widened daily experience.

Streetcars further widened that range of movement. Experiments with electric streetcars achieved practical success in 1887, when Frank Sprague's system—which installed separate electrical motors in each car so that they could be quickly assembled into trains of various lengths—proved capable of serving the hilly city of Richmond, Virginia. Horsecars were rapidly replaced after that, for they were slower, were more expensive to run, and dropped horse manure along their routes. Horsecar companies also had to maintain large stables, pay considerable veterinary fees, and risk epidemics in their stock. An electric line eliminated these problems, and the cars could be lighted at night and heated in winter. By 1890 no less than two hundred cities had built or ordered electric streetcar systems, which replaced most cable cars and the old horsecar lines. The electrics were faster, and the lines were extended into the countryside. Chicago, St. Louis, Pittsburgh, and other cities more than tripled their trackage.[61] The electric streetcar quickly became part of daily life, and in 1902 city dwellers traveled on it 4.8 billion times.

Streetcars offered new experiences of the city. Passengers were slightly detached from the passing scene, which became an unwinding panorama. Watching the spectacle framed by a moving window created an experience analogous to watching one of the new silent films. Each journey illustrated the fragmentation of experience. At the same time, riders learned to practice passive avoidance of the other passengers. They stared out the windows, read newspapers, or averted their gaze. As Trachtenberg noted, people on an urban streetcar were usually "distant, remote, and strange" to one another.[62]

These new forms of behavior were a contrast to the more personal encounters of pedestrians.

The energy transition affected behavior, as one could see with each new form of transportation. Streetcar passengers still shared a common experience in a social space, and they at least had the possibility of conversing, in contrast to those riding in automobiles, whose social distance from others increased to the point that it became a form of technological estrangement. Sitting behind windows, with their hearing temporarily impaired by the engine noise, unable to smell anything but the strongest odors that penetrated the metal container they sat in, motorists learned to accept sensory isolation from the world they traversed. The weak sense of community on the streetcar gave way to competitive individualism among motorists. Passengers on streetcars did not honk horns or shake their fists at one another, but motorists did. Aggressive drivers pushed ahead and endangered others, while poor or inexperienced drivers created difficult situations. States began to issue licenses to drive, but none required a driver's test until Rhode Island did so in 1908. Other states were slow to follow suit. Michigan, the center of the automotive industry, did not require a license until 1919.[63] The automobile had its pleasures, of course. It expanded the sense of bodily control and made possible new experiences of speed and maneuvering. Drivers were not confined to a fixed route, and they could spontaneously explore new places or stop wherever they wished. But along with the greater sense of control came status anxieties, based on the comparative quality and prestige of different cars. As Julie Wosk has noted, "The automobile, with all of its myriad social and psychological functions—its association with status and sexuality, enclosure and aggression—has become an extension of ourselves, a machine intimately connected with our sense of self-definition."[64] The shifts from walking to bicycling to streetcars to automobiles were not merely technical improvements. Each form of movement carried with it a different perception of the surrounding society and one's relation to it. In this process, people moved faster, encounters with strangers became anonymous, the sense of self became entangled with expensive technologies, and the experience of the world was gradually reduced to scenes framed by windows.

The civil society of the walking city was far different from that of the streetcar city, which centralized shopping in larger stores in the center, or the automotive city, which privatized movement and carried consumers away from downtown. The great energy transition not only accelerated movement but also redefined the relations between the individual and society. Neighborhoods became more differentiated along class lines, and individuals had less contact with one another. Increasing amounts of space were dedicated to driving, parking, and servicing automobiles. As Peter Baldwin explains, between 1910 and the early 1920s, "City people saw the car not just as a menace to life and limb, but also as an aggressor upon their time-honored rights to city streets."[65] Despite letters to the editor, protests, and lawsuits, however, arguments for automotive "freedom of movement" won out, and the streets that once belonged to pedestrians became noisy, polluted arteries of transportation. The number of automobiles increased from 6.7 million in 1919 to more than 27 million a decade later, with unanticipated consequences. Children no longer could play in the street. Noise and pollution from cars made a home's front rooms less enjoyable. It became less pleasant to sit on a front porch, and new houses often did not have one. Walkers or cyclists might converse with neighbors when passing, but drivers did not. The locations of shops and services gradually increased to automotive distances, neighborhood shops declined, and urban living became a bit more anonymous.

Americans used the great energy transition to transform their cities, to make radical increases in productivity through the mechanization of work, and to speed up the movement of both people and goods. They used new energy forms to disrupt labor markets, as some work disappeared, new kinds of employment developed, and workers changed jobs frequently. Many jobs were redesigned, and white-collar employment proliferated. More than twenty-one million homes and apartments were constructed between 1900 and 1929, the majority including the networked services of water, sewage, and electricity, and many older dwellings were retrofitted with these amenities.[66] During these transformations Americans redefined personal mobility, as they abandoned horses for bicycles, streetcars, and automobiles. Overall, the standard of living roughly doubled, a result that would have been impossible without the great energy transition.[67]

Politics did not drive these enormous changes or the new forms of behavior that came with them. The programs of the presidents did not explain the explosive growth of cities, the corresponding decline in the population and power of rural America, the disappearance of horses, the new job markets, the adoption of automobiles, or the skyscraper city. These changes were the result of an unprecedented increase in energy that made possible larger cities, faster transport, higher wages, larger profits, and a mushrooming gross national product. These trends created turbulence and uncertainty, preparing Americans to take seriously utopian and dystopian predictions about the future. Many political issues of the period, such as the prohibition of alcohol, the restriction of immigration, and the Ku Klux Klan's surge in popularity, were reactions to the growing dominance of prosperous cities over rural America. Prohibition was a "symbolic crusade" supported by "old stock" and rural Americans that was explicitly aimed at immigrants, who predominated in the fifteen largest American cities.[68] (Such matters will be discussed further in chapter 5.)

Not politics, but surging energy use explains the changing economy, the rapid growth of cities, the arrival and employment of millions of immigrants, the shortening of the workday, and the transformation of the nation's landscapes, discussed in the next chapter.

3

ENGINEERING NATURE

Thousands of engineers were needed to build the burgeoning cities, innovative factories, skyscrapers, and new infrastructure. At the Chicago World's Fair they built moving sidewalks, launched electric boats, and created spectacular illuminations At Buffalo's Pan-American Exposition in 1901, they constructed enormous models of their new factories. At the 1904 St. Louis Exposition models gave way to full-scale operations. The visitor could see a functioning cement factory, oil drills boring into the ground, and a thirteen-acre gulch that simulated coal mining. The most popular exhibit at the 1915 San Francisco Exposition was a Ford assembly line producing new Model Ts. Other exhibits included a working steel blast furnace and a gigantic model of the new Panama Canal. Collectively, the expositions represented progress through the work of inventors, engineers, and entrepreneurs, and the fairgrounds themselves were enclosed utopian landscapes of canals, towers, gardens, fountains, and vast exhibition halls. It seemed that nature could be sculpted and improved, which is exactly what engineers set out to do.

Engineers have seldom received much attention in histories of the Gilded Age and progressivism. Yet after 1876 the number of engineers in the United States exploded. As late as 1880 there were only 7,000; by 1920 there were 136,000. They had a central role in mining, irrigation, road building, urban development, manufacturing, and the great expositions. Engineering in the United States had begun as a branch of the military during the Revolutionary War. Later, courses were offered at West Point and other military schools. Until the Civil War, however, one often became an engineer through practical experience and apprenticeship on large projects such as canals, railroads, and bridges. As Terry Reynolds has documented, this hands-on training could be supplemented by engineering courses offered in the science and mathematics curricula at some colleges. Formal engineering education was also available at several academies, and by midcentury a handful of "polytechnics" had emerged, such as Rensselaer Polytechnic.[1]

Robert Friedel notes a number of other early engineering schools, and the expansion that came with the Morrill Land Grant College Act of 1862.[2] It allocated proceeds from the sale of federal land to fund agricultural and practical education, notably engineering. Each state could establish a new university or choose an existing one as its land grant institution. A few of those selected were private, such as Cornell and Clemson, but most were state universities, notably in Pennsylvania, Georgia, Illinois, Indiana, Iowa, Wisconsin, Michigan, Minnesota, Colorado, and California. It took a few years to recruit faculty and create the new engineering schools, but by the 1870s their first classes were being graduated. Early engineering education was mostly practical, but it gradually included more mathematics and science.

As the field professionalized, universities developed degree programs in mechanical, civil, electrical, and chemical engineering. The graduates joined professional societies that focused on power generation, illumination, transport, mining, construction, and manufacturing. This rising class of engineers played a central role in the great energy transition. They built roads, dams, bridges, and skyscrapers; developed new mining equipment; redesigned factory production; and mechanized agriculture. Many engineers became factory managers. They had to decide whether they identified with workers, with management, or with other engineers. Edward Layton's *The Revolt of the Engineers* examined those who defined themselves as progressives. In contrast, David F. Noble argued in *America by Design* that many engineers were co-opted into management. Others have studied how engineers attained a separate professional identity.[3] These are important discussions, but this chapter focuses on how the new cadre of engineers regarded and utilized nature.

Between 1876 and 1929 Americans had three different conceptions of nature. One aesthetic celebrated untouched nature and urged Americans to seek rejuvenation through contact with wild places. A second, pastoral aesthetic valued the fusion of nature and culture in gardens, parks, country houses, and family farms. This was a middle landscape between the urban and the wild. But the dominant aesthetic, and the one that most engineers preferred, was utilitarian, treating nature as a stock of raw materials. The natural world was to be transformed into high-energy landscapes of railroads, pipelines, bridges, power stations, factories, dams, canals, and other complex constructions.[4] This aesthetic culminated in the technological sublime, exemplified by enormous engineering projects and depicted in paintings and novels and described enthusiastically in political speeches.[5] The Lackawanna Viaduct over the Delaware River is a good example. When completed in 1910 it was the longest reinforced concrete structure in the world (figure 3.1).

As Cecelia Tichi explained in *Shifting Gears*, the engineers creating such sites became cultural heroes, part of a new elite that was based not on wealth, high culture, or tradition, but on technological skill. Between 1900 and 1930 the engineer appeared as the hero in more than a hundred silent films and in best-selling novels. He (seldom she) was "the representative man for the era, a symbol of efficiency, stability, functionalism, and power."[6] Engineers were dedicated to the application of mathematics and science to problem solving. As explained in a boy's book, *The Man of Tomorrow*, the engineer was "the man of action, the man who executes plans and put theories into effective practice." He was "the builder of our material progress. Through

Figure 3.1
Joseph Pennell, *The Lackawanna Viaduct*, etching, 1919. Courtesy of the Library of Congress, Prints and Photographs Division.

his ingenuity minerals are taken from the bowels of the earth and fashioned into instruments of production and transportation." Due to him, "great tumbling cataracts are harnessed and made to light our homes and to put into motion the intricate machinery of our factories." It concluded that without engineering, "our civilization could not be."[7]

In this utilitarian aesthetic, Americans were building a second creation that improved on original nature.[8] Industrial development was justified by narratives about using an axe to open the forest to farming, diverting a stream to drive a grist mill, building a railroad to open the West for settlement, or laying out an irrigation system to make the desert bloom. In each case, environmental transformation was narrated as though it had been less a human intervention than a natural development that occurred almost automatically. Build a sawmill, and the forest will be cleared away. Build a railroad, and settlers will people the countryside.[9] Such narratives assumed a partnership with the land. Forests were chopped down, rivers dammed, mines opened, and factories built to develop nature's latent powers. Development did not destroy the land but gave it full expression. Engineers were understood to be completing patterns sketched out in the "raw" nature of first creation, which was a rough draft that awaited full realization. Some canyons seemed to await a dam; certain mountain passes seemed providentially designed for a railroad. Second creation was understood to be not a violation of nature but rather its improvement and completion, in order to fulfill America's destiny. In these stories, the river wanted to turn a waterwheel rather than run uselessly to the sea; the prairie yearned for the plow so it could be productive; and the desert thirsted for water so it could bloom. President Theodore Roosevelt declared in his first message to Congress: "The western half of the United States would sustain a population greater than that of our whole country today, if the waters that now run waste were saved and used for irrigation."[10] Implicit in such texts was the message that the waters were ready to be impounded behind a dam and used. Each natural site had a higher purpose, and Americans had a duty to discover and realize it in partnership with the land. A book on irrigation argued in 1905, "The man who works intelligently in creating his irrigated farm with the raw materials of land and water, knows that in this smaller sphere he is engaged in finishing the world. He feels himself to be an instrument in the process of evolution."[11]

In this view, when Europeans discovered America, it was an undeveloped land that awaited their fructifying contact. The white settlers exploited resources that Native Americans had failed to use.[12] George Montague Wheeler, a West Point graduate who was first in his class in philosophy and mathematics and second in engineering, expressed a common explanation for why whites had a right to the lands in the West: "The ever-restless surge of population, almost a law unto itself, already in many cases crowd over the borders of these [Native American] reservations, and the time is not far distant when the question of the surrender of these lands to actual settlers will naturally be answered in the affirmative, on the plea of the greatest good for the greatest number." In the 1870s, Wheeler led a pioneering survey in the Southwest to identify sites for mining, lumbering, ranching, and irrigated farming. He believed in "wresting from productive nature the largest bounty," which Native Americans had not achieved but that "actual settlers" would.[13] Wheeler also was an entrepreneur in precision manufacturing, and one of the founders of the Elgin National Watch Company in 1864. It would eventually produce fifty million watches, including an expensive one bearing his name. Like many engineers, Wheeler focused on the management of both space and time.

Photographs of the Colorado River region taken on Wheeler's expeditions attracted national attention. At the same time, Thomas Moran made a celebrated painting of the Grand Canyon that was purchased by Congress and hung in Washington, DC. It celebrated the natural sublime, which fixed attention on the timeless or the eternal in nature.[14] But Wheeler and most other engineers were less interested in wilderness than in conquering space, mastering time, and overcoming gravity. Their attention shifted from nature to society, from constancy to change, and from metaphysics to technique. They were future oriented.[15] A sense of acceleration was fundamental to this aesthetic. The assembly line accelerated production. Bicycles moved faster than pedestrians; automobiles moved far faster than bicycles or horses, film conveyed a story more quickly than a novel, the amusement park packed more experience into a few hours than home amusements, and business decisions were accelerated by the telegraph and telephone. This quickening of life was linked to the energy transitions from muscle power and steam to gas, petroleum, and electricity. The new energy regime demanded a stupendous growth in the extraction of oil, coal, gas, copper, iron, limestone, and other raw materials. In all its forms—whether industrial production, construction, transportation, communications, or entertainment—acceleration increased the consumption of natural resources. Oil is an excellent example. Drilling for oil began in western Pennsylvania in 1859, and large deposits were later discovered in Texas and California. The early uses of oil were lighting, lubrication, and asphalt. Gasoline engines became important toward the end of the century. Yet well before then, fortunes were being made in the oil business, most spectacularly by John D. Rockefeller.[16] Oil was burned in some cities to produce electricity, and it eventually became an ingredient in the production of plastics, artificial leather, dyes, cosmetics, synthetic fibers, dentures, artificial limbs, and phonograph records. Even before 1900, when there were almost no automobiles, oil was an enormous industry. After that, it became gargantuan. Not only did automobiles, trucks, and buses need petroleum, but oil cut into coal's market for home heating. By the 1920s Americans owned a majority of the world's automobiles, and they extracted and burned most of the world's oil.

The Republican Party, which dominated national politics for much of this period, viewed the nation as a rich source of raw materials. The party platform of 1908 declared:

> The United States now owns one-fourth of the world's wealth and makes one-third of all modern manufactured products. In the great necessities of civilization, such as coal, the motive power of all activity; iron, the chief basis of all industry; cotton, the staple foundation of all fabrics; wheat, corn and all the agricultural products that feed mankind, America's supremacy is undisputed. And yet her great natural wealth has been scarcely touched. We have a vast domain of 3,000,000 square miles, literally bursting with latent treasure, still waiting the magic of capital and industry to be converted to the practical uses of mankind: a country rich in soil and climate, in the unharnessed energy of its rivers, and in all the varied products of the field, the forest, and the factory.[17]

As this statement suggests, the ideology of second creation treated nature as a warehouse of raw materials that ought to be used to improve the lifeworld. Nature was considered real estate to be improved. Swamps ought to be drained, and rivers were better dammed. Natural commodities were to be extracted from forests, lakes, mines, and quarries. A generation earlier the Transcendentalists had argued that human creation and natural creation were ultimately in pastoral harmony.[18] Such ideas were hard to sustain, however, when the railways tripled in length and entire forests had to be cut down to supply them with timber. In 1883 a federal report estimated that three million acres already had been cleared to supply railroad ties, and every year an additional 472,000 acres of trees would be needed for routine maintenance and replacement.[19] Likewise, demand for railroad rails and rolling stock stimulated steel production, which also expanded to supply the metal used in automobiles, skyscrapers, and myriad appliances, from toasters to refrigerators to hot water heaters. US steel production grew from 1.2 million tons in 1880 to more than 50 million tons in 1929, when the industry employed a workforce of 880,000.[20] The mass production of steel was possible because of the Bessemer process, which quickly removed impurities from the molten metal by blowing powerful blasts of air through it. Enormous mills were erected that relied on this process to make steel more inexpensively than had been possible before (figure 3.2).

Making steel required mountains of coal, and electrical power plants required yet more. Coal consumption grew rapidly, and its sales more than doubled between 1900 and 1929. Yet this growth was overshadowed by the explosive development of oil extraction, which grew from 25,000 barrels a year in 1876 to 2,670,000 barrels in 1929.[21] The following year for the first time the money spent on oil surpassed that spent on coal.[22] Well before then, the chief use of oil was for transportation, and drivers demanded better roads.

The first paved road was in New Jersey in 1870, decades before automobiles existed and even before bicycles became popular. Lobbying to improve roads occurred, but as late as 1900, only 4 percent of them had a hard surface.[23] In 1910, 75 percent of city streets were still dirt or gravel, and the 25 percent that were paved might be covered not with asphalt or cement, but with cobblestones, bricks, or even wooden blocks. Traditionally, householders had maintained and largely controlled residential streets, but they were seldom willing to bear the expense of a new pavement. When cities assumed the cost, they prioritized motorized traffic over other uses of the streets. After passage of the Federal Aid Road Act of 1916, more highways between cities began to be paved, on the condition that states matched federal grants.[24] During World War I the need for better roads became urgent, as military shipments were hindered by the poor condition of highways. After the conflict, in 1919 the US Army organized a cross-country tour that sent seventy-nine vehicles from Washington, DC, to San Francisco. It took them almost two months, and they averaged less than six miles per

Figure 3.2
Joseph Pennell, *On the Way to Bessemer*, etching and drypoint, 1908. Courtesy of the Library of Congress, Prints and Photographs Division.

hour. It conclusively demonstrated the need for better highways and for stronger bridges to replace the more than one hundred they damaged during the trip.

By 1929, all the states had a gasoline tax that established permanent funding for roads and stimulated demand for steamrollers, trucks, metal signs, concrete, asphalt, and gravel. Better roads encouraged drivers to go faster and further, with a corresponding rise in accidents. Higher speeds became more comfortable as roads improved, and as cars got better suspensions and more powerful motors. After 1915, millions of Americans acquired vehicles for the first time, and they enjoyed controlling such powerful machines and driving faster than a horse and buggy could go. The average automobile's speed roughly doubled from fifteen to thirty miles per hour between 1915 and 1930, with higher speeds on country roads than in the congested cities.

Road builders recognized the pleasures of motoring and began to design pastoral suburban parkways that offered new ways to experience the landscape. The Bronx River Parkway exemplified this development, as it transformed a cluttered and polluted river valley into a scenic drive. That parkway became famous, and other suburban roads copied its restriction of traffic to cars, banning horse-drawn carriages, trucks, and commercial vehicles. Parkways also prohibited billboards and small businesses along the roadside. They maintained an even traffic flow by replacing intersections with overpasses and using entrance and exit ramps to control access. These innovations ensured continual movement through a sculpted pastoral landscape.[25]

In a parallel development, Oregon's Columbia River Highway, built between 1913 and 1922, "established the state of the art for building scenic roads in mountainous areas."[26] Its tunnels appeared almost natural, with dramatic openings that framed views of the river. The rustic design of its guardrails and other safety features were integrated into the scene. Striking side trips were also included in its layout, such as hairpin turns in roads that snaked their way up to dramatic lookouts. Thomas Zeller explains in his history of automotive tourism that "the Columbia River Highway promised a new type of mobile scenic experience." The highway's promotors claimed that such roads "restored a connection to nature" and embellished the scenery. The *New York Times* concurred, calling the new road "a delightful thoroughfare" and "America's most noted example of man's intelligent development of nature's creation."[27] Similar techniques were used on roads in national parks. Highways were not merely transportation arteries. They orchestrated scenery to give motorists transformative experiences. They were not roads through nature but roads to nature, dramatic landscapes seen from a speeding automobile.

The automobile was not the only technology that Americans used to accelerate their experiences in nature. They adopted high-energy sports including roller-skating, bicycle racing, parachuting, and downhill skiing (made easier by mechanical lifts and towlines). Additional sports were invented. During the summer of 1922, using two boards for skis and a gasoline engine to power a boat, Ralph Samuelson invented water-skiing in Lake City, Minnesota, and it quickly became popular.[28] These sportspeople had a new relationship to the landscape. A generation earlier the hiker or amateur naturalist took a contemplative view of nature, seeking immersion in it. Automotive racing and the other new sports emphasized speed and a rush of impressions.[29] They emphasized vision, coordination, and balance, with less attention to the other senses or indeed to nature. The speedboat driver or the water-skier were not concerned with the aquatic ecosystem, and few automobile drivers worried about air pollution or their impact on roadside animals and plants. Increasingly, Americans raced through the natural world, treating the landscape as a backdrop to their self-dramatization. Their antebellum ancestors had inclined toward a romantic view of nature, but they wanted power over it.

This sense of domination and control was an essential feature of engineered landscapes of the emerging high-energy lifeworld. Their preeminent location was the city, with its skyscraper skyline, great bridges, electrical displays, impressive museums, and expansive scale. This high-energy landscape relied on motors and electricity that made it possible to run subways beneath the streets; transport customers effortlessly up and down the many floors of department stores; supply water, heat, light, and telephone service to offices and apartments far above the street; offer city residents a variety of foods and services; and stage a wide range of spectacular entertainments.

The mushrooming use of raw materials in cities, railroads, steel mills, factories, and automobiles created environmental problems: smoke pollution, a sanitation crisis, mountains of waste, and junkyards of derelict automobiles. As Martin Melosi summarizes, factories usually were located along major transportation routes near cities, and "modern factories—especially textile, chemical, and iron and steel plants—became major polluters." They often were located along waterways, which provided "the least expensive means of disposing of soluble or suspensible wastes such as phenol, benzene, toluene, arsenic, and naphtha."[30] New Jersey's Passaic River, once a commercial fishing ground, "died" from pollution in the late nineteenth century. In Pittsburgh, steel mills stained the rivers rust red, and in Cleveland the oil refineries killed aquatic life in the Cayuga River and polluted Lake Erie. "Meatpacking, which concentrated in cities such as Chicago and St. Louis, inundated adjacent areas with foul smells and dumped animal waste on vacant lots." Paper mills fouled the air and water with sulfur, chlorine, and other chemicals, and they demanded large supplies of wood, denuding surrounding forest areas.

However, because industries brought jobs to communities, industrial pollution was tolerated as a "nuisance," and dealt with haphazardly in response to complaints from residents. One sanitary expert felt that most city people took little interest in pollution, for "they look on the city as a sort of hotel, with the details of the management of which they have no desire to become acquainted." This was a classic example of the kaleidoscopic sensibility, in which the city was a spectacle to be observed. After circa 1900, however, Americans gradually recognized that pollution was a health hazard, and this widened to a general awareness of environmental issues. A class of mostly male professionals emerged who "developed systems to combat health hazards and pollutants, compiled statistics, and monitored some forms of pollution."[31] They curbed some of the worst excesses, and made successful reforms when allied with civic groups, often comprised primarily of women, whose clubs and associations pressured local authorities.

One example of this process was opposition to the coal smoke that polluted urban air before 1900. Smoke came from railroads, factories, and furnaces in homes and other buildings. Reformers campaigned against what they called "diseases of darkness" that were becoming endemic as cities became more crowded, tenements proliferated, coal smoke cast a pall across the sky, and ever taller buildings blotted out the sun. Sanitation specialists bolstered their cause with statistics on the health hazards of coal smoke. Its corrosive particulates blocked sunshine, harmed the lungs, and were linked to anemia, rickets, and tuberculosis.[32] The cures included smoke abatement legislation, zoning laws demanding setbacks on new skyscrapers (so more light reached the ground), teaching in outdoor classrooms, the adoption of sun lamps, and ingesting cod-liver oil or milk fortified with vitamin D. Just as important were the electrification of factories and urban railroads, which displaced the burning of coal away from urban centers to efficient power plants with tall chimneys, and the electrification of streetlighting, which reduced the burning of coal gas. Environmental reform succeeded when a scientific argument could be made against wasteful industrial practices, and an alternate technology or process could be substituted. Nevertheless, pollution remained widespread, and the era left behind noxious landfills, "dead" rivers, and vast urban dumps. As Melosi observed, especially during the nineteenth century, "economic as opposed to ecological considerations shaped the application of nuisance law in cases affecting the environment."[33]

Spills from early oil wells destroyed the surrounding countryside, and coal mining left behind tons of waste rock, waterlogged mineshafts, land subsidence, and accelerated soil erosion. The United States was the world's largest coal producer from 1890 until the 1950s, and the industry was particularly destructive in Pennsylvania, West Virginia, and Kentucky. Some fires in mines were not extinguished and continued to smolder underground for decades, and coal field wastes polluted nearby streams. Mining of iron, lead, nickel, copper, and other materials also led to widespread degradation, and in the worst cases made further use of the land impossible.

One such wasteland is Picher, Oklahoma, a once prosperous mining town that residents were forced to abandon. Picher was the site of extensive lead deposits. Before 1910 excavations were often so near the surface that two men and a horse could work a small mine.[34] Even as late as the 1920s 159 firms mined the 1,000-square-mile Tri-State area around Picher. However, the Picher Corporation acquired most of the richest ores, and it provided more than half the lead and zinc consumed by the United States during World War I. The core mining area covered approximately seventy-five square miles. In the 1920s the 14,000 men who worked in Picher's 1,400 mineshafts and 248 reduction mills were prone to illness and early death. The mines were dusty, and the reduction mills created piles of tailings hundreds of feet high that still tower over the abandoned town. When the Bureau of Mines studied the Picher district in 1923, it found that mine ventilation was poor and that many workers had silicosis or tuberculosis.[35] In response, the Bureau suggested methods to hold down the dust levels and urged that employees have medical exams every six months. However, the Bureau did not advocate compensation for workers who contracted silicosis or tuberculosis but allied itself with management. Over a five-year period ending in 1932, tests revealed that 6,108 miners had silicosis. Each one was immediately fired. Another study found that 62 percent of the workers who had been at Picher for nineteen years or more had silicosis.[36] Not only the men but the town was polluted. Lead constituted less than 10 percent of the ore and the town was surrounded by mountains of tailings that covered 7,000 acres. They contained not only lead but cadmium, iron, and zinc. When soaked by rain, the tailings became acidic and polluted the underlying aquifer.[37] The mines yielded billions of dollars in profits, but little was spent to deal with this waste or to retrain the miners for other jobs after the lead deposits ran out. Environmental hazards went unchecked, profits were invested elsewhere, sinkholes opened, tunnels collapsed, and the town shrank. Decades later it was declared a Superfund site where no one should live.[38] Picher is an extreme case, but it illustrates the problems created by coal, copper, nickel, iron, silver, and lead mines.

The abusive treatment of nature as raw material is further illustrated by the mechanization of mining. Early mines relied largely on muscle power for both extraction and bringing the ore to the surface. As mines penetrated deeper, however, flooding became a problem. During the early industrial revolution in Britain, the stream engine was improved to pump water out of mineshafts, and it became indispensable in many aspects of the coal industry. When shafts penetrated deeper, first steam engines and later electricity drove fans to improve ventilation, ran elevators, and pulled conveyor belts loaded with coal. Electric lights did not burn the scarce underground oxygen. When electric drills were introduced, output increased, but the number of miners did not. Mines shifted the funds previously spent on muscle power to the purchase of machinery.[39]

Greater energy intensity also made possible enormous open-pit mines. Mountain tops were removed to expose the ore-bearing strata that enormous steam shovels excavated. An open pit mine corkscrewed into the earth, creating a series of concentric shelves wide enough for the steam shovels and trucks to operate. The first pit, the idea of a MIT-educated engineer, was at the Santa Rita Mine in New Mexico, which began operations in 1910. The Bingham Mine in Utah adopted the practice. First it removed a mountain of copper ore and then spiraled downward. Driven by expanding electrification, demand for copper increased 600 percent between 1901 and 1929.[40] To produce it, copper ore was ground into sand so that the metal could be separated out, leaving millions of tons of waste. Smelting then spewed polluted smoke into the atmosphere. Every day at the Anaconda Copper Mine in Montana, despite having precipitators that recaptured two-thirds of the pollutants, the plant released twenty-five tons of arsenic into the atmosphere. In Hibbing, Minnesota, the largest open-pit iron mine in the world began as a conventional shaft and tunnel operation in 1895. It soon became an important supplier to US Steel. The sheer extent of the ore encouraged conversion to an open-pit mine like those in the copper industry. The Hibbing mine grew rapidly, and by 1919 it was ready to extract the ore beneath the town, which therefore was relocated. The mine still operates and has removed two billion tons of material, creating a six-square-mile hole that is called Minnesota's Grand Canyon.

Mines devasted their surroundings, but the most widespread geographical transformations of the energy transition were in agriculture. After the Civil War, settlers moved into the Great Plains, which become more arid as one travels west. The southern plains were largely devoted to cattle ranching, with vast herds driven to railheads each year. The plains were covered with grass, which could feed the cattle just as it had long fed the buffalo, before they were driven to near extinction on the southern plains by 1875. Ranching was also practiced on the northern plains, but farming became more common. In 1870, Nebraska, Kansas, and the Dakotas had 50,000 farms. By 1900 they had 400,000.[41] A thick sod of intermingled grasses protected the region's topsoil, before it was plowed up and transformed into a vast wheat belt. The pioneers were convinced that farming the Great Plains would improve the climate. There had been a serious drought in 1859–1860, and rainfall was irregular in some years. But between 1878 and 1887 more rain fell on the Plains, encouraging further migration. The popular expression "rain follows the plow" expressed this optimism, which helped railroads to sell land to immigrants and populate the countryside. Scientists at state universities documented the increase in rainfall, which suggested that plowing, planting trees as windbreaks, and modest amounts of irrigation were changing the climate.[42] This also seemed confirmed when the Mormons reported that the water level of the Great Salt Lake began to rise after they began irrigation. As John Opie summarizes, "The theory of increasing rainfall by plowing the land gained credence as a widespread scientific belief when it was endorsed by Joseph Henry of the Smithsonian Institution, the new American Association for the Advancement of Science, and the most famous explorer-scientist of the day, F. V. Hayden, director of the U.S. Geological Surveys of the Territories."[43] After 1887, however, rains failed and crops dwindled. The weather patterns proved to be cyclical, with moist decades followed by dry ones. Even with much more land than was needed to support a family in Indiana or Illinois, it was difficult to prosper on the high plains. After 1890, as one traveled west, the number of abandoned farms increased.[44] The idea that "rain follows the plow" turned out to be a fantasy of human control over nature, and the Great Plains became what agricultural historian David Danbom succinctly called "America's most disappointing frontier."[45] The settlers who remained had to master dry farming techniques, shift to winter wheat, or develop irrigation.

The railroads, gasoline-powered machinery, automobiles, and electricity transformed rural life. In 1876 the energy sources for agriculture were almost entirely the muscle power of oxen, horses, mules, and humans. The transition away from muscle power deeply affected rural life, even as the number of farms tripled between 1860 and 1910. At first muscle power was gradually replaced by steam engines that drove heavy equipment, such as threshing machines, and by the railroads that penetrated most rural areas, obviating the need to haul crops in wagons or drive cattle and hogs long distances to market. By circa 1910 gasoline motors were beginning to power tractors, cars, trucks, and agricultural equipment. By 1920 more than half of all midwestern farmers owned an automobile, and almost as many in the West did as well. Only one in seven farmers had a car in the technologically backward South. But overall, in 1920 rural America boasted more automobiles per family than urban America.[46] In 1930 more than 70 percent of all farmers outside the South owned a car.[47] Electricity from central stations was also unevenly distributed. It was rare in the southern states, but as Richard Hirsh notes, it had reached 40 percent of California's farms in 1925 and topped 50 percent in Utah and Washington by 1931.[48] As a direct consequence of acquiring new forms of energy, the equine population declined sharply after 1910. As horses disappeared, hours previously devoted to their care were available for other tasks, and the land previously used to feed them was released for other crops. Agricultural horsepower rose from 0.66 million in 1880 to 28.6 million in 1930. In that half-century, the productivity of wheat farming doubled.[49] Overall, farmers used tractors, new fertilizers, and electricity to increase their productivity by 50 percent.[50]

As machines replaced muscle power, the number of farmers and hired hands declined. The rural population peaked in 1915. Farmers comprised more than half the nation's population in 1860, but that figure fell to 44 percent in 1880 and 25 percent in 1930.[51] Mechanized horsepower changed the countryside's appearance. It worked best on large, flat fields, without hedgerows, and its efficiency encouraged mergers and the disappearance of smaller farms. As automobiles decreased the time needed to reach larger towns, crossroad communities shriveled. Farm families still had many children, but they often departed for the city. These changes were not imposed on a passive rural America. As Ronald Kline has demonstrated, farmers were active consumers. With considerable ingenuity, some used windmills, small hydroelectric dams, or car batteries to run small appliances and radios. They adapted gasoline engines to cut wood, pump water, and perform other tasks. As early as 1915, more than 90 percent of the farmers in Iowa had telephones, and their home furnishings often came from mail-order catalogues. By 1929 the farmer was deeply enmeshed in markets for energy, machinery, and consumer goods.

Mechanized energy had diverse consequences. Wendell Berry observed, "The kind and quantity of the energy we use determines the kind and quality of the life we live." The transition to gasoline-powered vehicles and electricity was accompanied by adoption of "the values of industrialization" in contrast to the more organic values of older family farms that were being consolidated into larger holdings.[52] In competition with larger, mechanized farms, homesteads "with small or steep fields are pushed farther and farther toward the economic margins and are finally abandoned." Heavy machines were efficient, but they also compacted soil that horse hooves had once churned up, leaving the land harder and less able to absorb rainfall. A farmer's degree of self-sufficiency declined as most gave up keeping a cow and no longer bred horses but bought machines. With little manure for their fields, they had to purchase fertilizers, often manufactured using the energy-demanding nitrogen fixation process.[53] Another new expense was hybrid seed corn, which was more productive than the farmer's own seed corn. It rapidly replaced open-pollinated corn after being introduced in the 1920s.[54] A farmer who tried to use part of a hybrid crop for seed corn discovered that the second generation was far less productive than the first, and therefore new seed corn had to be purchased every year. As self-sufficiency declined, the farm became more like a factory that continually paid for inputs of equipment, gasoline, electricity, seeds, and chemicals. Speed and efficiency became essential in planting, cultivating, and harvesting the increased acreage. Berry concluded that "the result is not farming; it is a process closely akin to mining."[55]

The industrialization of agriculture, the transformation of nature, and the decline of self-sufficiency were even more evident in the growth of irrigated farming, which scarcely existed in 1876 but had expanded enormously by 1929. At first, private irrigation was encouraged by the Desert Land Act of 1877. An individual could claim up to 640 arid acres for just twenty-five cents an acre if one were able to irrigate it, typically from a small stream or from a well fed by an underground aquifer. On parts of the Great Plains, one could drill down to the Ogallala Aquifer, which had been impounding water for thousands of years. Its 174,000 square miles underlay most of Nebraska; a strip of western Kansas, Oklahoma, and Texas; and a narrower strip of eastern Wyoming, Colorado, and New Mexico. In these areas some farmers could irrigate on an individual basis, although they withdrew water faster than the aquifer recharged.

Wherever there were rivers in the mountain west, large but expensive dams, pumping stations, and canals were possible. Some cooperative schemes mustered the financing and expertise necessary to do this, notably in Mormon communities. But it became clear that only the government had the resources needed to tame the larger rivers. This led to passage of the Newlands Reclamation Act of 1902 and the establishment of the Bureau of Reclamation, which President Roosevelt energetically supported. By 1917, the Bureau had completed twenty-one irrigation projects in the West including large canal systems that delivered water to otherwise dry land. The plan was to create a modern class of Jeffersonian yeomen farmers, tilling irrigated lands intensively on small holdings. In practice, however, irrigation water was so expensive that payments for it in the 1920s were millions of dollars in arrears.[56] Small farmers were squeezed out in favor of agribusiness. Increasingly, government water projects began to include hydroelectric generation so that dams could help pay for themselves. Irrigation exemplified the control of nature, but as with mining, corporations rather than individuals reaped many of the benefits.

Industrial agriculture was a matter not only of scale and expensive equipment, but also of regional specialization. There developed a corn belt, a wheat belt, and clearly defined areas growing cotton, tobacco, sugar, and rice. Dairy farming was concentrated in Wisconsin and Upstate New York.

By 1927 two-thirds of all canned fruit came from irrigated farms in California.[57] The conquest of nature increasingly meant industrial-scale agriculture. The semi-independent family farm in 1875 had horses and a cow or two, sold eggs to a local store, kept a large vegetable garden, had a small orchard, and harvested several cash crops. By 1929 such homesteads were disappearing.

Engineering nature at times demanded the wholesale expropriation of lands. Growing cities dammed distant rural valleys for water supplies. New Yorkers got some of their water ninety miles north of Manhattan. In the Catskill Mountains, between 1907 and 1915 towns were razed, a railroad line was relocated, and a dam and pipeline were built to impound 123 billion gallons of water in the Ashokan Reservoir. Bostonians were served by the Quabbin Reservoir, which displaced four small towns and many farmers. Hydroelectric dams also forced evacuation and flooding of Conowingo in Maryland and Rawsonville in Michigan. And everywhere, the electrical grid gradually expanded, creating regional interconnections between utilities. As Julie Cohn shows in her biography of the grid, it was not orchestrated by government, but cobbled together piece by piece, linking public and private institutions.[58] By 1929 there were regional networks organized around Philadelphia, Chicago, Denver, Boston, New York, and other cities, but a more comprehensive grid developed only later. The grid's construction at times appropriated Native American land for dams, transmission lines, mines, and roads. Some tribes were expelled from homelands that were designated as national parks, including Yosemite, Yellowstone, Glacier, and Redwood.[59] The displaced Native Americans became what Rob Nixon has called "uninhabitants." Their communities were "unimagined" through a process of "administered invisibility."[60] The parks, dams, reservoirs, and transmission lines were celebrated as though they had been established on vacant land.

Unimagined communities were not the only casualties of engineering nature. Many wild species declined, and some disappeared. Until the 1850s the "skies filled with clouds of waterfowl, with millions nesting in or migrating through Iowa." The marbled godwits appeared overhead as "vast rivers of birds," and there were thousands of prairie chickens that wintered in groups of three hundred. "Most numerous of all were passenger pigeons," which appeared in "waves of tens of thousands stretched from horizon to horizon" that took days to pass by.[61] Wagonloads of the birds were sent to market. This wild abundance declined as the settlers introduced European agriculture. Cows, chickens, horses, and other domestic animals replaced gophers, wild turkeys, prairie chickens, and quail, and drove them into small areas unsuited to farming. When wolves attacked sheep or when foxes raided the hen house, the predators were hunted down, often for a bounty. Wolf hunts were at times community events, where hunters surrounded a large area and gradually contracted toward the center, driving the wolves before them. Gray wolves were eliminated in most of the Midwest by 1885. The elimination of predator species was also federal policy in national parks. At the Grand Canyon, for example, between 1906 and 1930 the federal government killed 816 mountain lions, 863 bobcats, and more than 7,000 coyotes.[62] The Department of Agriculture established the Division of Predatory Animal and Rodent Control, which hunted, trapped, and poisoned mountain lions, wolves, coyotes, and bobcats. In Wyoming alone, it killed 40,000 predators.[63] At the same time, European invasive species, including thousands of wild burros, competed with native animals for the scarce vegetation. Invasive species of weeds accidentally spread across the land, notably the Canada thistle, garlic mustard, common buckthorn, Japanese knotweed, and purple loosestrife. The English or house sparrow was intentionally brought to America in 1852 and soon became one of the most common birds in the United States.[64]

Bison once had roamed in herds across the open plains, and their population before the region was transformed by agriculture was estimated to be twenty million or more. A mature bull weighed 2,000 pounds. Native Americans hunted the bison for meat and hides, found uses for their bones, and gathered their dung for fuel. In contrast, white hunters killed them for sport, and at times took only the tongues and the choicest cuts of meat, leaving their carcasses to rot. William Hornaday, one of the founders of the conservation movement, recalled in 1887: "They lived and moved as no other quadrupeds ever have, in great multitudes, like grand armies in review, covering scores of square miles at once. They were so numerous they frequently stopped boats in the rivers, threatened to overwhelm travelers on the plains, and in later years derailed locomotives and cars."[65] The railroads regarded the bison as a dangerous nuisance, and they hired hunters to eliminate them, including William Cody, later known as Buffalo Bill. The herds were also slaughtered for sport by visiting gentlemen hunters, in some cases encouraged and supported by the US Army.[66] By 1890 only 1,000 bison remained. Fortunately, some were protected in private herds, conservationists made efforts to save them, and they recovered, though they no longer roam the region they once dominated.

The decimation of predators, the extinction of the passenger pigeon, and the fate of the buffalo and many other native species worried some Americans. The nation was destroying irreplaceable parts of its environment. George Perkins Marsh sounded the alarm already in 1864 in *Man and Nature*, which documented the dangers of abusing natural resources.[67] He argued that the classical Mediterranean civilizations had caused their decline through deforestation and careless land use. Marsh's work inspired the next generation to attack wasteful industrial practices and to develop a program of resource conservation. Not only were watersheds protected and national forests created, but land management and forestry were transformed into scientific practices. President Theodore Roosevelt supported this conservation movement. He worried that the high consumption of wood would lead to a "timber famine," and he saw rivers as sources of power that ought to be put under control so they could be used for transportation, hydroelectric generation, flood control, and irrigation. These views were also endorsed by the major associations of engineers. They argued that "the potential electrical energy in the nation's streams should be harnessed to lessen the drain on exhaustible coal supplies. They promoted more efficient mining methods" and scientific management in factories. In their view, nature was being wasted if left alone. As one orator put it, conservation meant "a conscious and purposeful entering into control over nature through the natural resources for the benefit of mankind."[68] To this end, new agencies were established, including the Forest Service, the Reclamation Bureau, and the Geological Survey. As Samuel Hays notes, these agencies were part of a drive for national efficiency that "subordinated the aesthetic to the utilitarian. Preservation of natural scenery and historic sites," however desirable, were "subordinate to increasing industrial productivity."[69] The new specialists in forestry, for example, defined trees as a resource that ought to be grown, harvested, and replaced with seedlings, so that the next generation also had enough wood.[70]

Because Henry Ford's factories and motorcars made enormous demands on natural resources, he became interested in conservation. Ford appreciated the essays of Ralph Waldo Emerson, went to the Grand Canyon on vacation, and cultivated a friendship with the naturalist John Burroughs, with whom he once visited Thoreau's Walden Pond. Ford declared in his best-selling autobiography that the world was still "in the raw material stage," and he wanted to conserve resources.[71] He purchased forests in upstate Michigan, rubber plantations in Brazil, and other sources of raw materials, as well as a railroad to transport them to his vast River Rouge factory. He looked for ways to substitute alcohol for gasoline, to produce felt from agricultural waste, and to make plastic automobile components from soybeans. He declared, "We do not yet know how large a share of a modern automobile can be grown annually on the farm instead of exhausting the mines and forests."[72]

As a movement, conservation did not necessarily focus on preserving natural sites. Rather, the assumption was that to ensure material progress mankind should dominate and shape nature. Conservation was harmonized with the idea that the natural world was unfinished, awaiting a second creation. In this view, a settler improved land by clearing it, a dam improved a river, irrigation improved a desert, and a railroad raised land values by making a link to the market. This ideology of second creation sought to make an improved lifeworld. After 1870 the majority of railway track was laid; the skyscraper was invented in Chicago and spread to the nation's cities; huge mines extracted lead, copper, coal, iron ore, and other raw materials; impressive bridges spanned the Hudson, Niagara, Mississippi, and other rivers; massive hydroelectric dams were built; the Panama Canal was excavated; and millions of acres of dry land were irrigated in the West. Nature was further transformed by draining wetlands, expanding cities, and cutting across the countryside with transmission lines, pipelines, and paved highways. By 1930, the nation had been extensively engineered, and these changes were celebrated as a technological creation story. The great railway stations, skyscrapers, bridges, dams, highways, and airplanes seemed tangible proof of progress. In this view, untouched nature was a rough draft that could be perfected using new technologies. The railroad magnate James J. Hill once scoffed that "you might put a railroad in the Garden of Eden, and if there was nobody there but Adam and Eve, it would be a failure." Second creation required settlers to plow Eden and make it productive. Henry Ford declared on the final page of his autobiography, "We shall learn to be masters rather than servants of Nature."[73]

Much of Joseph Pennell's art was devoted to depicting the triumphs of second creation, including a series of images of New York, a series on the construction of the Panama Canal, and depictions of steel mills, factories, bridges, skyscrapers, viaducts, and other infrastructure. As one example, consider his *Steel Bridge at Niagara Falls*. Pennell focused on the great single arch of the bridge, which frames the factories that used the falling water for manufacturing. The famous Falls themselves are not depicted. He celebrated the control of nature to make it useful. In this era, monumental bridges and dams inspired powerful emotions: admiration for the skill of modern engineering, awe at the size and strength of the structures, patriotic pride, belief in progress, and a sense of humanity's increasing power (figures 3.3 and 3.4).

Figure 3.3
Joseph Pennell, *The Approaches to Gatun Lock, Panama Canal*, 1912, lithograph. Courtesy of the Library of Congress, Prints and Photographs Division.

Figure 3.4
Joseph Pennell, *The Steel Bridge at Niagara Falls*, 1910, lithograph. Courtesy of the Library of Congress, Prints and Photographs Division.

However, Americans did not always accept the engineering of nature. During the nineteenth century Niagara Falls had become the most popular natural tourist attraction in the country and a national icon. Nevertheless, by 1900 corporations had been granted rights to all the water flowing over the Falls, and the following year Buffalo's Pan-American Exposition celebrated the commercial uses of hydroelectricity. Visitors saw displays and exhibitions that highlighted its uses, not least in the brilliant lighting of the fairgrounds.[74] The exhibits educated visitors about the uses of the electricity. By 1908 one company was permitted to withdraw from Niagara 480,000 cubic feet of water per minute, and so many others had obtained water rights that the mighty cataract soon would be reduced to a trickle. The Colonial Dames of America, the American Civic Association, and many other organizations rallied against the industrial domination of Niagara, but businessmen and some scientists defended the utilitarian idea that, as *Scientific American* put it in 1911, Niagara ought "to work out its board and lodging" and provide its energy to society.[75]

Eventually a compromise was reached. The small tourist enterprises along the Niagara River, including fences that hid the view in order to sell tickets to lookouts, were cleared out and the area was landscaped, with free access for all. The industrial use of Niagara was rationed, with more water withdrawn at night than during the day. A sufficient flow was supplied to make the Falls seem natural to tourists, yet water was always being withdrawn for industry. One of the greatest attractions became the dramatic lighting of the Falls, giving them an entirely new appearance at night,[76] using not only white light but a rainbow of hues. Niagara was technologically controlled and presented in a new guise. The *New York Tribune* exclaimed, "Magnificently illuminated, the falls were of a beauty that their daylight aspect never equaled." As a bonus, "the garish outlines of the bleak brick [factory] buildings were gone," and the Falls were bathed "in a rush of prismatic color."[77] Niagara had escaped drying up, but it had been conquered and controlled in another way that vividly exemplified how the energy transition could be used to "improve" nature. Indeed, the Niagara electric power station itself became a tourist attraction.

Some spectacular natural sites were eradicated. The Tuolumne River in Hetch Hetchy Valley near Yosemite National Park was dammed to create a reservoir to provide water and electricity to San Francisco. Conservationists, led by John Muir and the Sierra Club, vehemently opposed its construction, but their efforts ended in defeat. In 1913, Congress passed the Raker Act, which allowed the valley to be submerged. Elsewhere, hydroelectric installations were often not opposed but celebrated as signs of progress, and many lakes behind dams became popular recreational sites. In a survey of dam sites Donald C. Jackson found that they were regarded as either monumental, like that at Hetch Hetchy, or pastoral. The thousands of smaller pastoral dams built between circa 1880 and 1930 "clearly involve a manipulation of the environment to serve a human purpose," but they could "meld into their agrarian setting in ways that do not detract from the surrounding flora and fauna."[78] Americans mailed and collected postcards of these smaller dams, and they clearly found them attractive and worth visiting.

In 1929, half a century of engineering nature seemed a success, whether one looked at the network of railroads, the rising productivity of agriculture, the domination of woodlands by professional foresters, the erection of hundreds of hydroelectric dams, the expanding grid of interconnected power lines, the vast open-pit mines, the skyscrapers that towered over American cities, or the millions of motorcars on the growing network of paved highways. Even the conservation movement focused less on controlling industry than on controlling nature. By adopting new forms of energy, the landscapes of the United States had been transformed, including far more than the expanding cities that now dominated the country. Utilizing new forms of energy, Americans had reshaped agriculture, penetrated the earth with mines and oil wells, reshaped rivers and watersheds with dams and canals, covered the countryside with a lattice of paved roads, edged those roads with electric transmission lines, brilliantly lighted the cities, conquered the skies with airplanes, and filled the air with radio broadcasts. It seemed the raw materials of nature were being reshaped to become a new lifeworld of second creation. There were problems, such as air and water pollution, but there seemed to be technological solutions. There were displaced "unimagined communities," but they were a little discussed "price of progress." Marchand noted a common theme in the media of these years, in which civilization was redeemed by new devices or processes: "The content of every medium of American popular culture of the era affirmed that man was constantly outwitting nature through technological advances."[79] The discomfort of driving on rough roads could be counterbalanced by superior springs and suspension. An over-the-counter drug could calm nerves overtaxed by the pace of modern life. A gentle laxative could overcome constipation caused by eating too many processed, "soft" foods. Pills could supply the vitamin D that once came from the sun, or restore the vigor lost due to the strain of overwork. It seemed the human body, like the natural world, could be improved and reshaped through an ever-expanding array of products.

American artists were fascinated by mechanization in all its forms, especially energy systems. Many painters and photographers depicted dams, airplanes, skyscrapers, and other technological achievements. The artistic movements of futurism, precisionism, and modernism built on the established American penchant for the technological sublime evident at the world's fairs. Industrial landscapes were celebrated in magazine articles, railroad advertising, postcards, poetry, and fiction.[80] James Oppenheim's 1912 novel, *The Olympian*, describes the view from a train window as it passed the factories outside Pittsburgh: "A vision shone and passed, swallowed in the night; the sublime spectacle of window-lit mills at the riverside, girdling the darkness with the fierce flaming of the Bessemer converter." Theodore Dreiser likewise felt exhilarated at the sight of an enormous steel plant outside Buffalo,[81] and many magazine articles celebrated the metropolitan corridor of factories along the railroads in the Northeast. Americans understood their canals, railroads, factories, bridges, dams, and cities as being not just practical achievements but symbols of national progress.

After circa 1900 artists painted and photographed both machines and their new social contexts. The canvases of realists such as John Sloan and other Ash Can School painters often expressed an awareness of the costs and ambiguities of technological change. In his early career, Sloan drew illustrations for radical magazines, and he understood the new electrified skyscrapers of New York as expressions of a shift in cultural power.[82] New York's skyline became an artistic subject, as architects, planners, and artists grappled with the new urban forms around them. The painter John Marin and the photographer Alfred Stieglitz depicted the Woolworth Building, and it was filmed by painter Charles Sheeler and photographer Paul Strand in 1921.[83] They recognized that they were living through a momentous transformation that these new human-made landscapes embodied.

This imagery was the diametrical opposite of the anti-modernism that sought refuge in the pre-industrial European past (discussed in chapter 6). The physician and poet William Carlos Williams considered T. S. Eliot's long poem *The Waste Land* a disastrous wrong turn. He wrote in his *Autobiography*, "I felt at once that it had set me back twenty years. . . . Eliot returned us to the classroom just at the moment when I felt we were on a point to escape to matters much closer to the essence of a new art form itself—rooted in the locality which should give it fruit." As an example of what Williams preferred, consider his poem from 1921, composed immediately after seeing a firetruck rushing through the streets in New York:

Among the rain
and lights
I saw the figure 5
in gold
on a red
firetruck
moving
tense
unheeded
to gong clangs
siren howls
and wheels rumbling
through the dark city.[84]

These lines inspired Charles Demuth to paint *I Saw the Figure Five in Gold* (1928), which hangs in the Metropolitan Museum of Art. It disassembles the firetruck into the headlights, the front axle, and vivid red and gold. Fragments of text—W.C.W., Bill, and Carlo—refer to Williams. Neither the poem nor the painting offers a narrative but rather an intense moment, detached from anything before or after. The same can be said about other poems by Williams or paintings by Demuth, including *My Egypt* (1927), which portrays light diffraction over an industrial building. Such works exemplify the precisionist movement that included George Ault, Elsie Driggs, and Charles Sheeler. They scrutinized the machines, factory districts, skyscrapers, and vibrant downtowns of urban America. Precisionist photography explored similar subjects, such as a Marmon automobile crankshaft (Paul Outerbridge, 1923), Ford's River Rouge factory (Charles Sheeler, 1927), Armco Steel (Edward Weston, 1922), New York City (Walker Evans, 1929), billboards (Ralph Steiner, 1924), and other industrial scenes.[85] These works often had such a tight focus that they might not be considered landscapes. They showed not a skyscraper or a skyline but rather the partial view of a building, not an entire factory complex but only a roof or smokestacks, not a car but one elegant wheel. The often-unspecified location made images almost abstract, shorn of human presence, geographical location, or narrative. These artists rejected both realism and impressionism, preferring to depict elegant machines or dynamic formal tensions. They took the kaleidoscopic impulse to a decontextualized extreme when they represented the technologies of the energy transition: generators, light standards, high tension lines, automobiles, airplanes, skyscrapers, radio towers, steel bridges, vast factories, grain elevators, or the symmetries of mass-produced machine parts, pipes, and steel beams.

Not all Americans celebrated the engineer as hero, the conquest of nature, or the modernist machine aesthetic. The subjugation of nature appeared rapacious to the historian James Truslow Adams. He had coined the phrase and defined "the American dream," but he also called the settlement of the West "raping a continent." In 1929 he declared, "It was not a question of preparing a continent for habitation. It was one of money-maddened men furiously wrenching wealth from it in every way their ingenuity and greed could devise, from the land, from the forests above it, from the mines below it."[86] Long before the Anthropocene had been named, Adams, Muir, and a few others found engineering nature alarming.

4

THE DISINTEGRATION OF TIME AND SPACE

The multiple effects of the energy transition were not limited to transformations in the workplace, the rapid growth of cities, faster forms of transportation, and the engineering of nature. Adapting to new energies altered the sense of time, space, and identity. The biblical sense of time, in which the world was created just a few thousand years ago, was disintegrating, replaced by millions of years of prehistory. The sense of space was challenged by the speed of travel and by technological innovations that made new environments possible, such as the subway, the skyscraper, and the amusement park. At the same time, the phonograph and motion pictures could record events, and the unique became repeatable. Overall, the individual experienced a paradox: Greater technical integration could evoke a feeling of disintegration.

This technical integration was epitomized by the linkage of the telegraph, the wire services, and the newspapers, which presented the national news every day. Before 1876 the flow of news had been episodic and mostly local, but with every decade it became more continual and national. Where once thousands of local editors in every city and town selected and interpreted the news to their local readers, increasingly the Associated Press and other wire services transmitted entire articles ready for use in syndicated newspapers. The flatbed press gave way to the rotary press after 1875, and setting type by hand began to disappear with the introduction of the linotype automatic typesetting machine in 1886. In the following year came the hot metal typesetting machine, which could compose and cast individual text at the same time. By the 1890s photographs were becoming common in newspapers, and after 1921 came wirephoto. News that had once been distributed by steamboats and railroads and printed by human muscle power was distributed through a vast electrified network and printed using steam engines or electrical motors. The modern newspaper carried both news and images from around the world. In 1924 Casper S. Yost, the president of the American Society of Newspaper Editors, declared in *The Principles of Journalism* that there was "no influence in the world so ubiquitous, so persuasive, so persistent as the newspaper." More than ever before, it presented "a continuous, never-ending moving picture of the world and its occurrences, of mankind and its conduct, depicting comedy, tragedy, vice, virtue, heroism, devotion, enterprise, discovery, calamity, beneficence, sorrow and joy—human life in all its kaleidoscopic and inexplicable changes." Readers in all parts of the nation received a shared, "kaleidoscopic view of varied events that unrolls before them as a continuous moving picture."[1]

Yet, paradoxically, this integrated system of news gathering and distribution contributed to the fragmentation of experience, exemplified by the newspaper's juxtaposition of unrelated and often distant events. On December 21, 1922, the editors of the *Miami Herald* reflected that "the front page of the morning paper presents the same kaleidoscope of action and emotion that is depicted on the screen or upon the stage." That day's front page contained "the story of a holdup, which in its boldness and its spectacular disregard for human life outrivals the James brothers," next to "a stirring story of mariners upon a great inland sea" who were "beaten by the tempests," next to a column on a cruel organized crime mob, next to a story about scientists seeking to produce electricity from wind-powered generators, next to the tale "of two boys who derailed a train in order to rob it," next to the account of an aviator carrying airmail who crashed in the mountains, and finally the mystery of a ship "careening along in the wind, without a living soul upon its decks." Each of these disparate front-page stories might have been the basis for a novel or a film. "Fiction can afford no more diverting mélange of action, or medley of emotion" than the newspaper's "cross section of life."

Far from criticizing this fragmentation of experience, the *Miami Herald* argued that "the daily round is prosaic enough," and the reader needed "excitements presented through the newspapers," books, and films. "These thrills are necessary in order to lift life out of the commonplace and to give it glamour."[2] But the typical potpourri of news had no coherence, and it arguably made the reader's life seem dull and uninteresting compared with thrilling events elsewhere. The fragmentation of information characteristic of the newspaper could also be seen in the newsreels shown in movie houses between feature films or heard in staccato news summaries broadcast on the radio. The mind became accustomed to headlines about far-flung, unrelated events. The structure of experience increasingly was shaped by a fragmentation of time, of space, of news, of entertainment, and of daily encounters with myriad unknown people.

The fragmentation of information was accompanied by a transformation of the sense of historical time. Until the middle of the nineteenth century most people believed that the world had been created just a few thousand years ago and that history had marched forward from that moment toward industrial civilization. Before geologists proved otherwise, it was widely accepted in Europe and the United States that the creation of the earth occurred immediately before the creation of mankind, who from the beginning were like the human beings of the present. The work of geologists destroyed this tidy scenario that put humankind in the center. They pushed back history millions of years and prepared the way for Darwin's studies, which concluded that human beings had slowly evolved from mammals. Scientists not only pushed back the history of the earth by millions of years but also helped to discover the mineral and coal deposits essential for the industrial revolution and for the great energy transition from water, wood, and muscle power to fossil fuels, steel, and electricity. The US government spent large sums on geological expeditions, and in 1879 it established within the Department of the Interior a permanent Geological Survey. Without geology, there could not have been such an energy transition.

All Americans benefited from geological discoveries, but many resisted evolutionary theory. Indeed, one-third of Americans still rejected it in 2025, preferring the "young earth theory" that posits that both the planet and humanity are about 10,000 years old.[3] This dispute concerns the relative importance of human beings in the universe. If humankind were present from the beginning, then the biblical story of Adam and Eve in the garden of Eden remains conceivable. But if human beings evolved from other creatures and only emerged after millions of years, then the biblical story is a mere fable, and human beings would be more accidental than divine. The radical expansion of time also undermined the fundamentalist belief in the second coming of Christ and the end of history. The Christian historical scenario was a focused drama with a beginning in innocence, a fall into sin, a savior, a battle for righteousness, and a spectacular Day of Judgment. Evolutionary theory proposed a vastly longer story in which humanity did not exist until near the end and had ignoble origins. This secular story stretched into a hazy future with no righteous conclusion but rather ended with the collapse of the sun and the heat death of the universe.[4] The Christian story was dramatically satisfying, and it included commandments on how to live a moral life. In contrast, evolutionary theory offered no uplifting guidelines. Little wonder that millions of Christians flocked to revival meetings that vociferously rejected Darwin. One of the most popular preachers was Billy Sunday, who in 1914 was considered one of the "ten greatest men in the United States."[5] He gave up a professional baseball career to upbraid sinners at revivals and offer them salvation. He regularly attacked evolutionary theory, alcohol, communism, and modernism. Christians who rejected evolution disseminated such views not only from the pulpit but also in publications, notably twelve booklets called *The Fundamentals*, which were first distributed between 1910 and 1915. An unintended result of these booklets was the invention of the term "fundamentalism" to describe this movement. The polarization between religion and science was to some degree also a geographical division. In the countryside, in small towns, and at religious colleges, the literal interpretation of the Bible dominated opinion. Some city people shared these fundamentalist views, but in the metropolis the theory of evolution was more generally accepted.

This conflict came to a head in the famous "Monkey Trial." In 1925 the state of Tennessee passed the Butler Act, which prohibited schools from teaching evolutionary theory. The American Civil Liberties Union decided to test the constitutionality of that law, in a case that involved John T. Scopes, a teacher in Dayton, Tennessee. He admitted to the new crime of teaching the theory of evolution at the local high school. His trial drew national attention when former secretary of state and presidential candidate William Jennings Bryan joined the prosecution. Bryan was a conservative Protestant who strongly believed that the Bible should not be criticized in the classroom. In response, Clarence Darrow, the most famous trial lawyer in the country and an agnostic, joined the defense. Given the clash of these famous personalities, the trial became a national event, covered by more than one hundred reporters and broadcast on the new medium of the radio. Scopes was convicted of having taught evolution and fined $100. On appeal, this fine was rescinded as being excessive, and Scopes went off to graduate school at the University of Chicago.

In urban America the so-called Monkey Trial was considered absurd. The newspaper columnist H. L. Mencken, who went to Dayton to report on the event, declared of the community: "Its people are simply unable to imagine a man who rejects the literal authority of the Bible. The most they can conjure up, straining until they are red in the face, is a man who is in error about the meaning of this or that text." Mencken skewered Bryan as well: "He can never be the peasants' President, but there is still a chance to be the peasants' Pope. He leads a new crusade, his bald head glistening, his face streaming with sweat, his chest heaving beneath his rumpled alpaca coat. One somehow pities him, despite his so palpable imbecilities. It is a tragedy, indeed, to begin life as a hero and to end it as a buffoon. But let no one, laughing at him, underestimate the magic that lies in his black, malignant eye, his frayed but still eloquent voice. He can shake and inflame these poor ignoramuses as no other man."[6]

The battle between Bryan and Darrow was more than a contest between evolutionary theory and the Bible. It pitted the city against small-town America. Fundamentalists enjoyed the economic benefits of geological knowledge, such as inexpensive gasoline and coal, but they refused to accept geology's religious implications. The Butler Act remained in force in Tennessee, and Arkansas and Mississippi passed similar laws. Much of the American South continued to oppose evolutionary science. In Dayton, the fundamentalists established Bryan College, which remains dedicated to teaching a "Christian worldview," which combines "understanding of the Bible and the liberal arts."[7]

If evolutionary time remained unacceptable to some, however, the value of being on time was widely accepted. The *McGuffey's Fifth Eclectic Reader* told schoolchildren that being on time was literally a matter of life and death. They learned that Napoleon lost the Battle of Waterloo and died a prisoner at St. Helena because "one of his marshals was behind time." Children were told that being late could cause a disaster. The *Reader* warned: a "railroad train was rushing along at almost lightning speed. A curve was just ahead, beyond which was a station where two trains usually met. The conductor was late,—so late that the period during which the up train was to wait had nearly elapsed; but he hoped yet to pass the curve safely. Suddenly a locomotive dashed into sight right ahead. In an instant there was a collision. A shriek, a shock, and fifty souls were in eternity; and all because an engineer had been behind time."[8]

By the last decades of the nineteenth century most Americans aspired to own a watch, both to regulate their own use of time and as a token of modernity. The enormous Sears, Roebuck mail-order company began business by selling watches. Anyone with a pocket watch quickly discerned that a town's clocks did not quite agree, and compared with the next town, clocks could diverge by anywhere from five minutes to half an hour. These inconsistencies made it difficult for people to be on time for appointments, for workers to arrive on time, and for passengers to make railroad connections. Hence the appeal of standardized time, harmonizing all the nation's clocks. Not the government but the railroads created a uniform system of time, which they put into effect on November 1, 1883. The railroads needed uniform timekeeping to synchronize their schedules and avoid accidents. Because it was convenient for most people, standardized time soon received widespread support, although few state governments required adherence to the new standard.

Alexis McCrossen has charted the rise and decline of public clocks, installed on churches, town halls, office buildings, railroad stations, and other prominent sites. These served as "markers of modernity." First at a local level and then, after the standardization of time at the national level, public clocks and bells divided time into shared units of experience and encouraged Americans to internalize mechanical rhythms. They replaced a sense of time based on observation of the sun and the moon, supplemented by tables published in almanacs. At the height of public-clock installation between 1871 and 1911, 15,000 were added to the thousands put up in previous decades.[9] People checked and adjusted their watches against these public timekeepers. As Michael O'Malley noted in his masterful history of American time, standard time "replaced nature and God with clocks and watches" and "changed the way Americans thought about themselves and their world."[10]

Being on time and using one's time efficiently became closely identified with both personal success and national progress. As Michael Adas found, during the nineteenth century Europeans and white Americans believed that they "had learned to measure, regulate, and make productive use of time to an extent inconceivable to any non-Western people."[11] McCrossen argued that modernity had been achieved "when time and space ceased to pose meaningful barriers to circulation of news, ideas, goods, and people."[12] Establishment of public time was crucial to breaking down these barriers, making it possible not only for railroad timetables to be harmonized but also for stock-market and commodities trading to be synchronized through the telegraph system. Moreover, as the electric grid developed, generating systems had to be synchronized in time and operate according to the same standards. Alternating current had to be harmonized at the same number of cycles per second before electricity could be interchanged between utility systems. Standardized time also was a prerequisite for shared experiences through the mass media. These linked all parts of the country to an occurrence as it happened, including presidential elections, sporting contests, battles, disasters, and legal confrontations like the Scopes Trial. Media events consolidated diverse publics into a single audience. Simultaneous experiences for the entire nation were impossible before time was synchronized. By 1929, such shared experiences were becoming a normal part of a new lifeworld. Technical synchronization enabled social simultaneity.

The federal government embraced standard time and enhanced its authority by erecting hundreds of clocks that adorned customhouses, courthouses, and post offices, as well as military installations. A close association developed between clocks and nationalism, and some sites proudly displayed four American time zones, usually represented by New York, Chicago, Denver, and San Francisco. The synchronization of the nation began to be expressed in monumental tower clocks, which included patriotic designs and figures such as George Washington. Americans "wanted to adorn their newly developing skylines with symbols of accurate time, national prestige, and imperial ambitions."[13] At the 1893 Chicago World's Fair, the US Navy exhibited an astronomical clock that sent an electric signal to 70,000 clocks all over the country. It was no longer acceptable for time to be roughly coordinated to within a few seconds. In the modern age, time became an absolute to be meticulously maintained, recorded, and used to coordinate a vast infrastructure.

Yet establishing time as an absolute was only one side of modernity. Despite its standardization, time could seem unstable, moving at different speeds in different situations. There was some resistance to standard time. Towns near one another but in different time zones seemed arbitrarily split. On one side the school day began in the dark, while a few miles away children started classes after dawn. One post office or bank opened or closed an hour earlier than the same services across that arbitrary line. Moreover, even as time became standardized and commerce moved more securely, some declared that the experience of time was not uniform. As Henry David Thoreau put it, time should not be "fretted by the ticking of the clock."[14] Mark Twain's neighbor in Hartford, the novelist Charles Dudley Warner, complained in *Harper's New Monthly Magazine* in 1884 that "the chopping up of time into rigid periods is an invasion of freedom, and makes no allowances for differences in temperament and feeling."[15] If industrial time was mechanically measured out, by contrast, natural time was seasonal, cyclical, irregular, and flowing, with occasional moments of great intensity.

A kaleidoscope can present transformations so rapidly that the eye can scarcely follow them, or it can be turned slowly, displaying the evolution of new patterns, as demonstrated by the new sport of synchronized swimming. It emerged after Annette Kellerman performed an underwater ballet at the New York Hippodrome in 1907 and in several aquatic-themed silent films. In 1917 she appeared in another spectacular show at the Hippodrome with "a supporting cast of 200 mermaids."[16] Synchronized swimming became part of physical education at the University of Wisconsin, and it became popular at Ohio State University by the 1920s. The dancing movements of synchronized swimming emphasized symmetrical patterns of apparently identical bodies that moved precisely in harmony with music. At the 1934 Century of Progress Exposition in Chicago, sixty swimmers moved to the beat of the twelve-piece band. Similar entertainment proved popular on the stage, notably at New York's Radio City Music Hall, where chorus girls wearing identical costumes danced in synchronized patterns, as each woman in the line kicked her legs, swung from side to side, and gestured as one. Their mechanized gyrations suggested the orchestrated movements of mass production. When Siegfried Kracauer saw such a performance, he concluded that the identity of the female performers had been submerged in a spectacle that made them into parts of an erotic assembly line.[17] Both synchronized swimming and chorus lines objectified a woman's body as an interchangeable part and celebrated conformity to mechanization.

It is common in communications studies to point out that the telegraph and telephone annihilated space and time. It seems indisputable that the experience of time was fundamentally altered by instantaneity. The sense of the present, rather than being located within one's sight and hearing, expanded to include anywhere that could send a signal. Before circa 1848 newspapers carried stories from other parts of the country that were weeks old. In the second half of the nineteenth century, the speed of news transmission accelerated, especially during the dramatic events of the Civil War. By the 1870s, the telegraph system reached all parts of the nation, and it had become normal to hear the news immediately. When a dam broke and a wall of water destroyed Johnstown, Pennsylvania, on May 31, 1889, the catastrophic death of 2,209 people instantly became national news. The following day the disaster was reported not only in the eastern and midwestern papers; it was also the lead story in Los Angeles, Sacramento, Portland, Salt Lake City, and Cheyenne.[18] As news became national, local stories diminished in importance, and Americans began to share a simultaneous experience of historical events.

In *The Culture of Space and Time* Stephen Kern devotes over a hundred pages to the discussion of time between 1880 and the end of World War I. In reaction to the establishment of a single public time and the increasing bureaucratization of the period, he argues, Victorians chose to "affirm the reality of private time against that of a single public time and to define its nature as heterogeneous, fluid, and reversible."[19] Moreover, public time, even though defined as a constant, forward-moving progression, seemed to be accelerating. By 1900 Henry Adams believed that the pace of history was increasing geometrically.[20] Saving time and increasing efficiency contributed to a torrent of events seemingly beyond human control.

If one could standardize time and accelerate events, traveling through time also began to seem plausible. In the 1880s Mark Twain's *A Connecticut Yankee in King Arthur's Court* sent its protagonist more than a thousand years into the past, and Edward Bellamy's *Looking Backward* projected its narrator a century into the future. Neither Twain nor Bellamy made any attempt to conceptualize time travel in scientific terms. But in 1895 H. G. Wells conceived of a time machine that could take people forward or backward to whatever moment they chose, and other authors also explored this theme.[21] If Christian fundamentalists struggled to retain the belief that the world was only 10,000 years old, stories of time travel adopted an expansive sense of time from evolutionary theory. Time travel fiction was merely speculative, but Albert Einstein's theory of relativity was soon understood to suggest that if one traveled fast enough in outer space, time would slow down. Science fiction stories began to appear in which space travelers did not age at the same rate as people on earth. Paradoxically, time was becoming both more precise and relative.

The mastery of space was also a hallmark of kaleidoscopic America. Industrial society celebrated the triumph over space, first in the expansion of railroads across the American continent and after 1903 in the conquest of the skies through aviation. In 1870 the United States had 52,922 miles of railway track, including one line to California. The system was already long enough to go twice around the world. In the next twenty years, it tripled in length, to 163,597 miles. The system included thousands of bridges, viaducts, trestles, and tunnels, which testified to enormous investments and the ingenuity of the railroad engineers. Descriptions of railway journeys in these years often contrasted the travails of early pioneers in covered wagons with the ease of rushing across vast distances while asleep, reading newspapers, smoking a cigar, or enjoying a meal in the dining car. Mark Twain described in *Roughing It* the hardships he endured in 1861 when taking a stagecoach through the Great Plains, where passengers were thrown about and battered by dislodged luggage and a heavy dictionary. A decade later, the new transcontinental railroad offered an experience of ease and luxury. As recounted in a newspaper story, a party departed from Omaha and soon "sped along at the rate of thirty miles an hour." They "agreed it was the fastest living we had ever experienced. (We beat that, however, two days afterward when we made twenty-seven miles in twenty-seven minutes, while our champagne glasses filled to the brim spilled not a drop!)"[22]

Journeys became even more convenient once the railroad corporations standardized the width of the track, at a 4-foot, 8.5-inch gauge.[23] This made all rolling stock interchangeable. Passengers had to disembark and move to another train less frequently, and less cargo had to be unloaded from one car and reloaded into another. This reduced the time, the cost, and the labor needed to ship goods. Every engine or car could traverse the entire country, including the luxury railway cars that belonged to wealthy families. These rolling apartments, complete with kitchen, sleeping quarters, a lounge, and office space, allowed wealthy families to roam across the nation without, in a sense, leaving home.

As the railroad system linked the nation more tightly together, it had many social and economic effects. Economic specialization by region began before the Civil War and became more pronounced after 1870. Railroads enlarged the market for perishable goods, which could be carried to all parts of the country before they spoiled. Refrigerated railway cars made it possible to slaughter animals in Chicago or Omaha and sell the meat nationwide. Consumers adjusted to buying vegetables from California, citrus fruit from Florida, cheese from Wisconsin, and other goods from far-off locales. Any factory with access to the railroad network could ship its products nationwide. The enlarged markets created by railroads helped corporations achieve economies of scale, which increased profits. Companies such as General Electric, the Ford Motor Company, or Sears, Roebuck could rationalize production and shipping by placing their facilities along major rail lines. Sears located its headquarters in Chicago because it was served by many railroads. From there it shipped prefabricated houses, cast-iron stoves, furniture, books, clothing, and tools to all parts of the country. In Detroit, Ford bought a railroad that interconnected with all the major lines running through Michigan and used it both to obtain the raw materials needed to make automobiles and to ship parts to assembly plants around the country. General Electric's factories were located on major rail lines, notably in Schenectady, New York; Erie, Pennsylvania; Cleveland, Ohio; and Fort Wayne, Indiana.

These new factories exemplified an extreme level of the organization and domination of space. By the 1920s it was not unusual to move river channels and flatten hills before erecting a complex of buildings laid out to maximize production on an assembly line. Such factories became popular tourist sites, where the public could witness the subjugation of natural forces and the organization of energy systems at a scale that strained the imagination.[24] The Panama Canal, completed in 1914, was another example of the domination of space. To build it required blasting away the sides of mountains, creating dams and reservoirs, installing a system of giant locks, and the electrification of the entire facility, including electric locomotives that pulled ocean liners through the canal, and extensive lighting so the locks could operate day and night.[25]

The comprehensive mastery of space was particularly evident in the city. Boston expanded by filling in a large shallow back bay, where elegant brownstone homes, churches, hotels, and a magnificent new public library were erected. Chicago reversed the flow of its river and raised the ground level of the downtown. New York broadened Manhattan Island with landfills and linked its boroughs with tunnels, bridges, and ferries. Americans created new kinds of urban space that could be seen from the elevated railroads, the summits of skyscrapers, the walkways of enormous bridges, in subway tunnels beneath the streets, in vast sports stadiums, and in cavernous railway stations. These were sites of the technological sublime that tourists sought out and that city residents celebrated as landmarks of modernity.[26] At the same time, mastery of plate glass manufacturing made it possible for the walls of steel-framed buildings to be more transparent.

Beginning during the 1890s in Chicago and New York, and soon after in every American city, the facades of buildings contained more glass. Windows in houses and apartments also grew larger.

These new spaces were brilliantly lighted, whether the factory, the street, store interiors visible through the plate glass windows, or the towering walls of skyscrapers. Electricity did not merely replace gaslight; it transformed the appearance of the entire city. Where there had been one space, there were now several versions of that space. The urban dusk and night, when electrified, restaged every cityscape, letting the imagination soar. Lights fringed the boulevards with receding streaks of flame. Illuminated ships sailed by on the rivers, light bulbs outlined the great bridges, and the railways and streetcars rolled bright windows across the night. Electricity illuminated the clocks, powered giant advertising signs, turned public buildings into stately landmarks, and remade every aspect of urban space. At different times of day, in varying weather, the city's scenes continually shifted, making the metropolis into a kaleidoscope of views, always itself but ever changing.[27] After 1911 the invention of the motograph spelled out running text in advertising signs. Other inventions made it easier to simulate moving images, and by the 1920s a sign had to be animated in order to compete with the energetic gyrations of other signs nearby.[28]

An entire generation of painters and photographers was enthralled by these scenes. They recognized that the electrified city represented a break in the continuity of human experience. Before 1910 artists such as John Sloan, J. Alden Weir, and Childe Hassan painted urban twilight scenes. As Hassan noted, "in the early evening when just a few flickering lights are seen here and there, the city is a magical evocation of blended strength and mystery."[29] After circa 1910, urban lighting intensified further, with more powerful arc lights on the streets, additional lighting in store windows, larger electric signs, and floodlighting on the skyscrapers and public buildings. Painters began to focus on this brilliant landscape after sundown, with its striking contrasts, for example, Georgia O'Keeffe in her *New York, Night, 1929*. Light edited the city to contain only illuminated magnificence. All that was shabby or smacked of poverty faded away into blank, apparently empty space. The strong contrasts and brilliant colors of the night city drew the attention of Joseph Stella, who deployed his American form of Futurism in the five canvases *The Voice of the City of New York Interpreted*. Two of these paintings depict the Great White Way, which, in Stella's words, "churn with dazzling splendor: they look like monstrous wings of fantastic butterflies. Broadway—the White Way—noisy and shrill, wild and aflame with pleasure seeking."[30]

Especially for visitors from small towns or the countryside, walking in the electrified city was thrilling. In 1915, one visitor to New York recalled:

> To walk Broadway after the theaters have disgorged on Saturday evening is to enter a romantic land where night is not, and gayety is unceasing. Electric signs dazzle on every hand and men and women good-naturedly jostle each other in the stream that washes the sidewalks. Overhead, in a blaze of light, a huge eagle with a fluttering ribbon caught in his beak flaps his nightly journey toward a five-foot bottle of beer, a kitten tangles and untangles herself in a spool of well-known silk, and a huge name of cigarette fame lends its luster to the dazzling highway. From a tall building, Broadway seems a long curving ribbon of white light.[31]

Those more familiar with Great White Way considered it to be the apotheosis of advertising. Waldo Frank declared in 1919, "The hard dazzle of Broadway by night is the very antithesis of gayety. It is business. Broadway's brightness is the sum of advertisements flashing petticoats and constipation-cures against the blackened heavens."[32]

Businesses had also changed in response to the compressed and heightened experience of space and time. The telegraph accelerated the flow of information and consequently the pace of decision-making. Information distributed at the speed of stock tickers and the telephone demanded constant attention. The scale and the speed of production increased. Electrified factories could run additional shifts after dark, and in some cases, they ran around the clock. These changes imparted a dynamism and uncertainty to life that was both exhilarating and disorienting. Already by the 1880s, the acceleration of all aspects of life seemed to tax the nerves, and in the following decades the pace of business relentlessly increased.

In 1881 George Beard's best-seller *American Nervousness* argued that rapid technological change exhausted city dwellers. Modernity seemed to threaten Americans with psychological stress. Beard believed that nervous exhaustion had become especially severe in the northeastern United States, then the most industrialized and urbanized region in the country. The "chief and primary cause of this development and very rapid increase of nervousness is modern civilization, which is distinguished from the ancient by these five characteristics: steam power, the periodical press, the telegraph, the sciences, and the mental activity of women."[33] Beard explained the growing prevalence of nervous exhaustion by comparing the human body to a steam engine. "The nervous system of man is the center of the nerve-force supplying all the organs of the body. Like the steam engine, its force is limited." To some extent, that force could be "increased or diminished by good or evil influences, medical or hygienic, or by the natural evolutions—growth, disease and decline; but none the less it is limited; and when new functions are interposed in the circuit, as modern civilization is constantly requiring us to do, there comes a period," no matter how strong the individual, "when the amount of force is insufficient to keep all the lamps actively burning; those that are weakest go out entirely, or, as more frequently happens, burn faint and feebly."[34] It followed from this analogy that the pressure to be on time was a burden on the nervous system. As Beard explained, the "perfection of clocks and the invention of watches have something to do with modern nervousness, since they compel us to be on time, and excite the habit of looking to see the exact moment, so as not to be late for trains or appointments. Before the general use of these instruments of precision in time, there was a wider margin for all appointments." Before time was imposed as a tyranny, "men judged of the time by probabilities, by looking at the sun, and needed not, as a rule, to be nervous about the loss of a moment, and had incomparably fewer experiences wherein a delay of a few moments might destroy the hopes of a lifetime." The conclusion could only be: "Punctuality is a greater thief of nervous force than is procrastination of time. We are under constant strain, mostly unconscious, oftentimes in sleeping as well as in waking hours, to get somewhere or do something at some definite moment."[35] While Beard's study of nervousness is no longer an accepted scientific work, psychologists still propound theories to diagnose the stress and strain of modern life.

Beard described Thomas Edison as "one of the healthiest men I ever knew," one who could make demands on himself that would break down the health of most people. Yet, already in 1881, "The experiments, inventions, and discoveries of Edison alone have made and are now making constant and exhausting draughts on the nervous forces of America and Europe."[36] Beard wrote these lines when there were scarcely any electric lights in Manhattan, no electric advertising, no electric streetcars, and no cinema.

Beard was not alone in developing new metaphors for the human body based on new forms of energy. When muscle power was dominant, a person was often compared to a horse. A young woman might be called a filly;

a young man, a stallion. An elderly woman might be termed an old nag. Someone who resisted control might be called skittish, while an indefatigable laborer was a "workhorse." There were also popular expressions such as "getting the bit between your teeth," being "off his feed," or having "a burr under his saddle." Steam engines inspired other metaphors, such as "getting up a head of steam" or being "under a lot of pressure." In 1860 a newspaper article entitled "The Human Coalburner" declared that "every man is not only a locomotive but a fuel consumer too,"[37] and went on to extol the efficiency of the human body as a machine. But in the emerging high-energy lifeworld human beings, instead of being compared to domestic animals or steam engines, were gradually reconceived in terms of wires and motors as electrification spread through society. People began to use such expressions as "getting your wires crossed," "an electrifying performance," or suffering a "mental short circuit." The brain was increasingly understood as an electrical system that had affinities with a battery that could be run down and recharged. Such ideas were expressed formally in the new science of ergonomics, which defined human beings as inefficient machines that needed training to maximize their potential. E. J. Marey argued, "As we regulate the use of machines in order to obtain a useful result with the least exertion of work, so man can regulate his movements . . . with the least fatigue possible."[38] Similar ideas inspired Frederick Winslow Taylor and other scientific managers, who imposed on the bodies of workers their vision of rationality and efficiency. The consumption of new forms of energy had reciprocal effects on consciousness.[39] Not only did the consumer continually adopt new metaphors to understand the self, but managers demanded that the bodies of workers conform to "rationalized" routines.

Even as human beings were being reconceived as machines rather than as animals, experience itself was being translated into moving images. The first films were short loops showing something as simple as an acrobat juggling. The medium advanced rapidly to tell short stories, to report the news, and, twenty years after its inception, to feature full-length films. As cinema matured, directors learned how to manipulate space and time, to create special effects, and to juxtapose images. Film audiences learned to accept conventions of movement, as cameras tracked action from one room to another. They acceded to simulations of events, to their abbreviation, and to sudden shifts in time and place. As the illusions of film became more complex, it was possible to represent in one hour a story that unfolded over several years, or even decades. By 1915 the pace of experience in film was far faster than in everyday life. These experiences taught the audience to perceive the world in new visual ways, as they became accustomed to sudden shifts in perspective, to juxtapositions, to collage, and to other breaks in continuity. The kaleidoscope itself had helped prepare viewers for such visual innovations. Discontinuity was being normalized.

At the same time, the public was challenged to accept a new understanding of sound. Until Edison's phonograph, invented in 1877, sound was ephemeral. There was no way to preserve or listen to anything more than once. One had to be there and be attentive when the soprano sang, the thunder struck, the orator spoke, or the crowd cheered. For about a decade after the phonograph appeared, people paid admission to theaters to hear demonstrations of the new wonder.[40] Many thought it must be a hoax. Often these doubters were asked to sing or say something that was recorded and then immediately played back. Audiences were asked to try to tell the difference between a live voice and a recording. Even if they could do so, such demonstrations seemed to demonstrate the fidelity of the new technology. Yet this apparent fidelity could be manipulated as recording technology became more sophisticated. Producers and singers discovered that microphones recorded nuances that audiences could not hear in a concert venue. The microphone made possible studio recordings that featured intimacy and subtle details that could not be heard by a crowd in an auditorium. Just as cinema developed the close-up that revealed delicate expressions on an actor's face that were invisible in theatrical performance, the microphone could isolate and amplify nuances of an individual voice.[41] Furthermore, the use of multiple microphones allowed studio engineers to balance the sounds of all the performers in an ensemble, and after the recording session they could make refinements. The final product was a carefully doctored sound that could not be reproduced in a live concert. Once the phonograph severed a sound from its time (and separated voice from physical presence), the result was both the preservation and the transformation of sounds (and soon after, images) into products. Like time and space, they were increasingly splintered, manipulated, and commodified.

Between 1876 and 1930, these changes in experience were on display at world's fairs and amusement parks. At the top of the Eiffel Tower two wax figures still sit in a small glassed-in room, where they have been for more than a century. One is of Gustave Eiffel, whose firm specialized in steel bridges before he designed and built the tower as the centerpiece of the 1889 Paris Exposition. The other figure is that of Thomas Edison, whose electrical inventions had been prominent at world's fairs for the previous decade. Both men were famous for practical achievements. The Eiffel Tower was extensively lighted using both gas and electricity. Elsewhere on the exposition grounds, the Edison company had a display that was fifteen meters high. It employed 25,000 colored electric lights to simulate the French and American flags waving in the wind, while flashing the words "Paris" and "Edison."[42] One might say that already in 1889 Edison was virtually present at several locations.

The fragmentation of identity, the mastery of space, the acceleration of time, and the manipulation of images and sounds were all on display at world's fairs, where two contradictory manifestations of the kaleidoscopic metaphor confronted one another. The progressive vision was evident in the educational exhibit buildings erected at the center of each fairground. One reporter at the opening of the Philadelphia Centennial Exposition of 1876 felt so overwhelmed by the scale of the spectacle—"head whizzing with all sorts of sights and sounds"—that it seemed impossible to capture in words: "it would be like attempting to describe the colors and figures of a kaleidoscope." He declared that "in the life of an ordinary man, there are few such days in any part of the world" and that "no man living on the continent of America can reasonably expect to see exactly another, or even one remotely approaching it, as a grand, imposing, and magnificent spectacle."[43]

A reporter from Baltimore was equally dazzled by "handsome fountains, rich hangings, and other decorations; innumerable banners and trophies and superb collections . . . [including] rich collections of decorated furniture, glittering chandeliers, groupings of richly colored woolen, silk, and cotton goods, beautiful carpets, rare wood carvings, and the most beautiful mosaics. On entering this immense kaleidoscope of tints, one is at a loss to know where to turn, so endless, apparently, is the vista of gleaming colors."[44] The displays included inventions and refinements that in their variety suggested rapid advances in every aspect of life. At each subsequent fair, the exhibits in the manufacturing buildings alone were so vast that no visitor could view them all. If one included the enormous halls devoted to electricity, mining, agriculture, art, and indeed every branch of knowledge, each exposition evoked a characteristic comment from fair visitors: "We were on the grounds most of seven days, and we certainly left more unseen than we had time or opportunity to examine."[45]

One popular entertainment at both the Chicago and Omaha expositions was Buffalo Bill's Wild West show. In 1898 the *Emporia Gazette* reprinted an ecstatic review of Buffalo Bill's "Wild West and Congress of Rough Riders of the World": "'Imagine,' says the *New York Sun*, 'a kaleidoscope, with an object field four and a half acres in extent, occupied by a swiftly-moving mass of figures, individually picturesque, brilliant with metallic reflections and gay with colors, momentarily springing and flashing into new combinations and modes of motion, which dazzle, confuse, and fascinate the eye of the beholder. Such is the spectacle presented by the hundreds of 'rough riders' forming part of the Wild West show, when going through their evolutions together."[46]

Buffalo Bill's show became kaleidoscopic through rapid action. To make expositions resemble the view through a kaleidoscope, the entire scene was thrown into motion each evening by spectacular illuminations. Beginning in 1881, expositions were among the first sites to be extensively lighted, at a time when many towns still relied on gaslight and only a few private residences had electricity. Almost all brilliant illumination was then experienced in public spaces, such as central squares, department stores, and railway stations. Early expositions literally displayed more electric light than many Americans had seen anywhere before, especially visitors from rural areas. These lighting effects instantaneously transformed the scene. Space itself suddenly seemed malleable, as the color and shape of every building changed, not once but several times in a few minutes, and every transformation was reflected in an exposition's canals, pools, and artificial lakes. Elaborate lighting displays were among the most popular features of the expositions in Chicago (1893), Omaha (1898), Buffalo (1901), St. Louis (1904), and San Francisco (1915), each of which surpassed its predecessors with ever more spectacular effects.[47]

One of the most popular forms of illumination was the electrical fountain. Already in 1884 one of these had attracted crowds at the International Electrical Exposition in Philadelphia. Its indoor fountain with fifteen jets of water periodically became the center of attention when all other lights were extinguished. Colored bulbs and filters enlivened the display. By 1889, the Paris Exposition had greatly improved the technique, and a "nightly show of illuminated fountains entranced crowds with a spectacle of falling rainbows, cascading jewels, and flaming liquids."[48] In 1893, the fountains at Chicago's Columbian Exposition were even more mesmerizing. Visitors enjoyed two large electric fountains at the western end of the Grand Basin, which was surrounded by the largest buildings of the fair. These fountains had 340 jets that sprayed 22,000 gallons of water into the air every minute, and as darkness fell, they were illuminated by powerful electric lights hidden below the water. An operator continually changed the fountain's colors, using filters that could make the water ruby red, sea green, yellow, blue, or various combinations of these hues to include "every color that the prism refracts." Nozzles divided the water into different configurations or made it pulsate or shoot suddenly higher, at times surging 150 feet into the air. These kaleidoscopic variations were reflected in the enormous basin.[49] On warm evenings, thousands lingered to watch the fountains, while they listened to music played by a brass band. After the Chicago Exposition, similar fountains were at the 1898 exposition in Omaha and erected in many cities[50] (figure 4.1).

Evenings at the Chicago fair ended with spectacular fireworks supplied by Paine's Fireworks from New York City. These displays cost as much as $10,000 a night, a stupendous sum at that time. At first, they were held around the great basin, but so many attended that they were moved to the

Figure 4.1
Electric fountains, Omaha Exposition, 1898. Stereo photograph, Strohmeyer & Wyman, Publishers. Courtesy of the Library of Congress, Prints and Photographs Division.

shore of Lake Michigan, where a launching platform was constructed out on the water. The elaborate fireworks were reinforced by powerful searchlights that shot bolts of light into the sky. On July 4, 1893, more than 200,000 people witnessed the display, which began with the spectacular launch of a giant balloon, which rose high above the crowd. "There was a flash of light, a shower of sparks, and the American flag was revealed in brilliant flame suspended in midair." This flag was as large as a football field and was composed of hanging chains covered with red, white, and blue fireworks. Suspended beneath a barely visible parachute, it slowly floated down and winked out, replaced by a bombardment of rockets and bombs bursting with stars. Below, on the waters were "Roman candles, golden fountains, fiery geysers," and other special effects that shot out of the lake.[51] One observer noted, "It was something gorgeous and deafening while it lasted, and it lasted an hour. At one time the lake shore for half a mile was fringed with intense magnesium fires, which threw a white, unnatural light over the dense, immovable thousands. When the people looked at the fires, the press of humanity had a dark, somber shade. When a rocket went up and the faces were lifted, a sudden pallor spread itself over the great area, a weird transformation, seen only by those who looked down from the lofty rail of the peristyle. Chicago has seen fireworks before, but never anything like those of the Fourth."[52]

In contrast to the progressive integration of the formal, educational exhibits, an exposition's amusement zone offered visitors disorienting experiences, whether in the roller coaster's terrifying plunge, the funhouse distorting mirrors, the whirling swings, or other physically harrowing rides. The only purpose of such amusements was to provide a good time. In the 1890s, amusement parks opened across the country, often owned by street railways. They were typically located outside towns and cities, often beside a lake, at the edge of a forest, or on the seaside. As John Kasson noted, in these places white people mingled from all ranks of society. Hierarchies broke down, and many formalities of public life were suspended. The "rides were designed to throw people off balance, literally and imaginatively. . . . [They] served in effect as powerful hallucinogens, altering visitor's perceptions."[53] Electrification was essential to much of the entertainment, which often used motors, lighting, tracks, and other streetcar technologies.

At a fair's educational exhibits, the visitor was given uplifting information. One was expected to read captions, listen to explanations, and gaze in fascination. The visitor was pensive and somewhat passive. But in amusement zones the visitor was thrown into action. One guidebook shrewdly observed that the "most popular entertainment to be found in any great exposition or county fair consists in those features which make the spectator a part of the show." He praised the amusement zone at the San Francisco Exposition of 1915 because "there were more places to ride, more places to frivol, more bumps to bump and more scenic treats underground and through mid-air, than had ever before been offered a show-going public."[54] These amusements typically included a maze inside a house of mirrors that shattered normal perceptions of space and perspective, and a roller coaster that threw riders into the sky, cast them down steep declivities, whipped them through terrifying curves, and altogether disrupted their posture and composure. After screaming through what seemed a near-death experience, one might wander into a "fun house" where distorting mirrors suddenly made one tall and slim, short and fat, or grotesque. The amusements pushed the limits of sound, speed, and perception. They seemed to defy gravity, distort and blur vision, upend control over the body, and empty the mind of everything but extremes. They were engines of disorientation.

A visit to the amusement park accelerated the pace of events, packing a huge number of experiences into a few hours. This compression of experience was also a central feature of the reenactments of disasters at the parks. These included the volcanic eruption that buried alive the inhabitants of ancient Pompeii; the destruction of Galveston, Texas, during a hurricane; and the cataclysmic Johnstown Flood that swept away a Pennsylvanian city when a dam collapsed upstream.[55] Luna Park on Coney Island staged "the fall of the Turkish city Adrianople in 'the first Balkan War' of 1912–1913."[56] Fantasies of destruction were also common in dystopian novels and short stories at the turn of the century, including Ignatius Donnelly's *Caesar's Column*, in which "a foul and brutal and ravenous multitude" sacks and burns New York City. Its destruction was repeatedly imagined between 1870 and 1930, as Max Page analyzes in *The City's End*.[57] There were stories of invasion from abroad, of mass death due to a freak accident or the poisonous gas from a comet's tail, or of vicious class conflict that ended in chaos. In contrast to the narrative of technological progress, there lurked in the popular imagination fears of sudden, comprehensive destruction.

At amusement parks, visitors could view an enthralling disaster and then throw themselves into the whirling motion of spinning machines that they could not control. There was the giant rotating barrel, a horizontal cylinder that challenged one to remain standing while walking through its turning. Halfway through this "barrel of fun," one crossed a threshold into a second cylinder spinning in the opposite direction. Anyone who retained their balance might still fall after colliding with someone in front of or behind them. Many people ended up on their knees, crawling and sliding toward the exit. They had ceased to be spectators and become part of the spectacle. Once on solid ground again, they could look back and watch others lose their equilibrium.

Another common amusement park ride consisted of a large turntable of polished wood. People climbed on and tried to stay aboard as it spun faster and faster, until all had been flung off, to slide across a polished floor and come to rest on the upward sloping sides of the room. Reginald Marsh's painting *Human Roulette Wheel* depicts this ride, which threw "bodies into intimate contact, creating chance romantic encounters, intense disorientation, physical entanglement, and voyeuristic thrills for spectators."[58] At almost every park, the swings were another popular ride. Patented in 1904 as the Traver Circle Swing,[59] it was designed "to provide the occupants with the pleasurable sensation of the ordinary roundabout with increased speed and gradual rise from and return to the ground." This movement resembled the experience of flying, and thrilled millions of riders at parks across the country. As the speed of rotation increased, the swings no longer hung straight down, but rather were flung outward, and the seat rose higher in its flight. As the rotations increased, the objects of the fairgrounds gradually became a blur of colors and lights, a visual disintegration.

Towering over the Chicago Fair's amusement zone stood another rotating ride, the first Ferris wheel (figure 4.2). It was erected as a private venture at the cost of half a million dollars, with no state subsidy, and it fully paid for itself by selling tickets for fifty cents. The Ferris wheel could carry 6,300 people every hour, and even running half-full it made $1,600 an hour. It was a huge revolving wheel, with thirty-six railroad cars suspended on its circumference. The twenty-minute ride consisted of two revolutions. Each time the wheel stopped, six cars were emptied and filled with new passengers. Every car carried as many as sixty passengers to its highest point 260 feet above the grounds, commanding a view of the exposition, Lake Michigan, and downtown Chicago. Unlike the static view from a skyscraper, the Ferris wheel provided a dynamic experience of movement.[60] The ride became more prominent after dark, because the great wheel's rim was studded with "three thousand incandescent lights of various colors" that were

Figure 4.2
Ferris wheel at the Chicago Exposition, 1893, Waterman Co., Chicago, Illinois. Courtesy of the Library of Congress, Prints and Photographs Division.

"arranged in groups on the rods and around the crown of the wheel. These are alternately extinguished and re-lighted at night as the wheel revolves, giving it the appearance of a monster display of pyrotechnics."[61] The Ferris wheel after dark afforded a fine view of the illuminated fairgrounds and the lively amusement zone. Even Henry Adams greatly enjoyed the ride.[62] After the success of the Ferris wheel in Chicago, copies were erected at amusement parks across the country.

These parks underwent a rapid evolution. In the 1890s "the average park was patronized by the rough and tough element," and "unprotected ladies and children dared not go." But amusement parks were soon made respectable, with security forces employed to maintain good order. By 1905, they were larger and offered a wide range of activities, usually including a beach and picnic grounds, as well as rides. They had become respectable places "where any person can go with perfect propriety and safety," as one street railway manager noted.[63] Albert Bigelow Paine described "The New Coney Island" in *Century Magazine*. He contrasted the low life and petty crime of a decade earlier with the amusement park's transformation into a haven of pleasure. "We were not surprised at the multitude of entertainments, but at the improved quality of them, and at the general deportment and evident respectability of the pleasure-seeking visitors"[64] (figure 4.3).

The nation's most famous amusement parks were on Coney Island, attracting prodigious crowds, who came by bicycle, boat, and streetcar. In 1905 the Coney Island Street Railway carried 39.5 million people to the parks. The parks were also seen in silent films produced by Edison's Manufacturing Company. Those made just after 1900 depicted daytime activities, including a young couple strolling through the attractions, children wading, various rides, and a famous diving horse. In 1905, Edwin S. Porter filmed *Coney Island at Night*.[65] For three minutes it slowly panned over Luna Park and Steeplechase Park, providing a panoramic view of the spectacular lighting. To make it, Porter had to rejig his camera to make time exposures that could capture the illuminations but not the tumultuous movement of rides. And because film was then silent, it could not record the boisterous crowds or riotous sounds. Coney Island also attracted artists, drawn by its exuberance and fantastic lighting.

At the great world's fairs, the contrast between the uplift of the educational exhibits and the perilous pleasures of the amusement zone expressed the root contradictions of the kaleidoscopic metaphor. The exhibits emphasized progress. They surveyed knowledge and achievements, celebrated new inventions, and embraced cultural cosmopolitanism, all of which pointed to a progressive, rational future. This kaleidoscope was firmly under control. It turned slowly, and visitors studied an unfolding pageant of possibilities. In contrast, on the midway visitors were inside the kaleidoscope, immersed in frenzied sensations, exoticism, hedonistic pleasures, and they vicariously enjoyed disasters, with no thought for tomorrow. As Fred Thompson, the impresario who owned Luna Park, put it, "Your average person . . . wants a conception to dazzle, to stun him, to hit him on the head like the crack of doom."[66] Amusements overwhelmed visitors, who threw off restraint

Figure 4.3
Louis Lozowick, *Coney Island (Luna Park)*, 1925 lithograph. Courtesy of the Amon Carter Museum of American Art, Fort Worth, Texas.

and capitulated to immediate sensations. Yet despite these differences, the educational exhibits and the amusement zone shared some characteristics. Each was an intricately designed technological realm, organized to handle large crowds. Each packaged experiences into discrete units for rapid consumption. Each appealed to a wide public among the white population but usually excluded African Americans and other groups. Each was a temporary construction that would soon disappear. Together, they expressed the tensions of an emerging lifeworld, including its hopes, fears, exclusions, disorientations, pleasures, fantasies of control, and bedazzlement.

If one were to single out one machine that epitomized both control and dangerous immersion, one machine that aroused the greatest wonder and enthusiasm of the age, it was the airplane. In the last decades of the nineteenth century, inventors struggled to invent a heavier than air machine that could rise into the sky and navigate the heavens. So many well-financed attempts failed, however, that there was little press coverage of the first successful flight, made by two bicycle mechanics from Dayton, Ohio, on a remote Carolina beach in the fall of 1903. The Wright Brothers were outsiders, without government support or university degrees. But they were meticulous workmen and practical analysts. They invented a wind tunnel where they could test and improve prototypes, using models to discover what worked best. They developed a new shape for the propeller that did not mimic fans or boat propellers but rather imparted both forward motion and an upward lift.[67] They jealously guarded their innovations, making few public demonstrations before they had secured patent rights. While few Americans may have heard about their exploits in 1903, the Wrights soon became famous, notably at public demonstrations in France and in Virginia in 1908. When they first flew over a river, they tied a canoe to the bottom of the plane in case they were forced to land unexpectedly (figure 4.4).

When a plane flew over Chicago in 1910 an estimated one million people rushed into the streets to see it. Later, in Washington, DC, the highlight of Independence Day celebrations was the overflight of twenty airplanes.[68] Until the 1920s the sight of a flying machine could bring out a crowd, and aviators were hired by state fairs to perform. Few people had ever flown themselves, however, and for decades newspapers continued to publish accounts of what it felt like. In 1916, a small Iowa newspaper reprinted a story that described the wonders of flying in a Zeppelin airship:

> The preeminent sensation of flying is buoyancy—you are a soap bubble in the air—you are swimming again on inflated bladders as when a boy in the old swimming hole—you have lost gravitation—weight is annihilated—your fear is not that you may drop to earth, but that you are never going to alight again—you have become a satellite to Mother Earth and you don't care a darn. The air you breath is aerated champagne. . . . You gaze down as through a most fascinating kaleidoscope on the world and for the first time realize what a "bird's eye view" really is. You see an entire city at once. In one glance you see what it

Figure 4.4
Wilbur Wright examines a canoe attachment to the airplane before the brothers' first flight over water, September 29, 1909. Courtesy of the Library of Congress, Prints and Photographs Division.

> took a thousand years to build. It looks like the blueprint of an architect, only the buildings are not drawings they are real. The river is not a crooked line, it is running water. The trees are real trees and that little man on the funny, narrow street is alive. An octagonal skyscraper looks like a brick wall pulled out of its hole. You see the heart of a solid city block and find pleasant gardens and trees and a fountain in the hidden court. The ruined castle perched on a mountain peak, green with ivy and moss, with its lookout tower high above the trees, shows its loveliness.[69]

Another reporter, Catherine Oglesby, summarized her first airplane flight for a Texas newspaper in 1919, which included a storm, a forced landing, and a nosedive into a field: "the physical sensations and the mental visions which visit one on one's first ride in an airplane are a kaleidoscope of colorful change."[70]

At the amusement park, mechanical rides rushed visitors down roller coasters, heaved them into the air on Ferris wheels, and disoriented their senses on rides. But in the end, they returned safely to the ground. Airplanes offered a more extreme experience, for they were not tethered to the earth but soared in the unpredictable air and disappeared into clouds, producing the unprecedented sensations of a new lifeworld in the sky. Conquering gravity, the pilot seemed a pioneer of the future, experiencing the world in entirely new ways, soaring, swooping, looping, turning, and diving. That future was potentially terrifying during World War I when airplanes became weapons. In peacetime, flight suggested freedom and the exploration of new realms, but in battle it portended aerial combat, attacks on cities, and bombing far behind front lines. The conquest of the air was exhilarating, yet fraught with danger. America seemed to accelerate into the unknown.

II

REACTIONS

5

PROGRESSIVISM AND BOUNDARY MAINTENANCE

The lifeworld of 1929 was a sharp contrast to the lifeworld of 1875, especially if one looked at rural and small-town America, where the majority of the population lived at the centennial of independence. Unless one took a train, it was a slow, muscle-powered life, sharply divided into day and a dimly lit night. The work week was over sixty hours. Life expectancy was about forty-five years. Most people lived in close proximity to useful domestic animals: horses for transport or to pull agricultural equipment, cats to combat mice and rats, often a cow for its milk, perhaps chickens for their eggs, or a dog to go hunting with. The horizon, the news, and even the time of day was local, not national. If people wanted entertainment or music, that too was mostly local, though an occasional circus or lecturer came through. Information was from printed media or conversations. There were no telephones, phonographs, loudspeakers, radios, or motion pictures. This existence was upset by the high-energy society that rapidly emerged after 1875, and the change provoked strong reactions.

The transformations discussed in part I created new conditions and suggested the possibility of radical change. Chapter 2 examined how the energy transition facilitated the sudden expansion of cities, the reorganization of work, and the adoption of new forms of transport, all of which inspired both utopian hopes and apocalyptic fears. Chapter 3 discussed the creation of the engineered landscapes of industrial agriculture and irrigation, the transformation of mining, the erection of vast new infrastructures of transport and communication, and the efforts to eliminate smoke pollution, improve public illumination, pave streets, excavate sewers, build mass transit systems, and in other ways create "the city beautiful." The many world's fairs discussed in chapter 4 showcased new forms of architecture and entertainment that the energy transition made possible. All of these innovations infused Progressivism.

One can understand many reforms of the era in terms of the five phases of the great energy transition to oil, gas, and electricity. In the 1880s, when the new energies were in their first stages, many transformations seemed possible, including the utopia described in Bellamy's *Looking Backward*. By the 1890s, as cities continued to mushroom, pragmatism became more evident. In 1890 Jacob Riis's *How the Other Half Lives* drew attention to the desperate need for better sanitation and housing reform, and he ended that book by describing superior new tenements recently built in in Brooklyn. He concluded that "model tenements can be built successfully and made to pay in New York, if the owner will be content with the five or six per cent" return on investment."[1] Thus in the second phase utopian ideas gave way to practical demonstrations, whether of better tenements, new streetcar systems, or improved street lighting. Such new infrastructure began to be built in the third phase, when almost every city installed these technologies. The crucial phase of technological momentum for the new energy forms arrived in the decade after 1910, when many roads were paved; when cars, trucks, and tractors replaced horses; and when electricity saturated urban space. The transition culminated in the fifth, mature phase after World War I. During the 1920s most homes were rapidly electrified, automobile ownership became the norm, and factory productivity soared on new assembly lines. As these technologies were adopted, both utopian hopes and apocalyptic fears faded.

Other kinds of reform followed a similar sequence of events. Extreme expectations during early years were followed by more pragmatic results. The Grangers and Knights of Labor declined, and the American Federation of Labor grew stronger. Grassroots movements were consolidated in the Populist Party of 1892 and then absorbed by the Democratic Party. In this process, radical ideas, such as nationalizing the railroads, disappeared. But more modest reforms succeeded, such as the direct election of senators by voters rather than their selection by state legislators. Likewise, reformers got the referendum that enabled voters to propose and pass new laws, and the right to hold special elections to recall an official.[2] These changes made American democracy more inclusive.

During the Roosevelt and the Taft presidencies breaking up corporate monopolies became a Republican policy.[3] In practice, however, Roosevelt regulated the great corporations far more than he abolished them. After journalist Ida Tarbell exposed the corruption of John D. Rockefeller's Standard Oil,[4] it was dismembered into separate oil companies. But when

General Electric was investigated and discovered to own several ostensible competitors, the company was fined but allowed to absorb them. Progressives defined the gas, water, and electrical utilities as "natural monopolies," because it seemed senseless for a city to have two or more sets of electrical wires, water pipes, or gas lines. But since there was just one utility for a given service, it also made sense to control prices and ensure good service by creating state regulatory boards.[5] These efforts did not return the United States to a nation of small businesses; they increased regulation. However, a federal department of energy was not established until 1977.

It might seem that business interests were on the defensive and forced to concede much of their power. But the work of Gabriel Kolko and other historians shows that many business leaders supported progressive reforms because they accepted the consolidation of corporate power and because they were watered down in implementation.[6] Regulation was seldom punitive, and corporations preferred it to the nationalization called for by socialists. Roosevelt and Taft were pragmatists, and the laws they enacted were compromises that business accepted. Even the income tax was initially so low that it was almost symbolic.

Progressives also reformed the corrupt administration of American cities. It might seem that civil service reform was not related to energy, but the frantic construction boom during the great energy transition created a bonanza in bribes and kickbacks. Political appointees readily accepted bribes for issuing building permits, rights of way for mass transit, or contracts to install streetlights, the pavement of roads, and park landscaping. The installation of telephone lines and exchanges, gas pipes, fire alarm boxes, sewer systems, fire hydrants, and water lines demanded thousands of workers, engineers, investors, suppliers, and administrators, as well as regulators and safety inspectors. Every service could be a source of kickbacks, including garbage disposal, street cleaning, tree trimming, lawn mowing, or washing the windows of public buildings. The rapidity of urban growth created a potential bonanza of corruption.

"Muckraking" journalists aroused the public to action. In *The Shame of the Cities*, Lincoln Steffens described how "St. Louis exemplified boodle [bribery]; Minneapolis, police graft; Pittsburgh, a political and industrial machine; and Philadelphia, general civic corruption," making it the most corrupt city in the country. In St. Louis, "Franchises worth millions were granted without one cent of cash to the city, and with provision for only the smallest future payment; several companies which refused to pay blackmail had to leave; citizens were robbed more and more boldly; pay-rolls were padded with the names of non-existent persons; work on public improvements was neglected," and the funds were diverted to elected officials.[7] In Minneapolis, a mayor corrupted the police force, which, in exchange for a share of the profits, turned a blind eye to gambling, prostitution, and organized theft. In Pittsburgh, a mayor built a strong political machine that even included members of the supposed opposition. Steffens concluded that the "graft of Pittsburgh falls conveniently into four classes: franchises, public contracts, vice, and public funds," all of which the political machine controlled.[8] Philadelphia had the same corrupt practices as other cities, and it had so perfected voting fraud that nonexistent people, deceased citizens, and even dogs were registered voters who perpetuated the machine's power.[9] Chicago also had colossal problems. Its police force was too small, and it was "first in violence, deepest in dirt; loud, lawless, unlovely, ill-smelling," and polluted, and yet it had managed some reforms.

Corruption was inseparable from the "spoils system." The winners of an election had the right to fill positions throughout the city with their supporters. A job candidate's qualifications often were less important than party loyalty. As a result, municipal services might be poorly run and serve the citizens only to the extent that they benefited the party in power. The reform of the civil service began in 1883 with the federal government's passage of the Pendleton Act. It created a Civil Service Commission to administer competitive examinations as the basis for a merit-based appointment system. The political parties could no longer assess officeholders for campaign contributions, nor was refusal to contribute ground for dismissal. At first, the act only covered 10 percent of the roughly 100,000 federal employees, but it was soon extended to post offices and customhouses. By 1895 Theodore Roosevelt, then commissioner of the Civil Service, could declare that "fifty thousand places are now taken out of the political arena. They can no longer be scrambled for in a struggle as ignoble and brutal as the strife of pirates over plunder; they no longer serve as a vast bribery chest with which to debauch the voters of the country."[10] This reform weakened political machines and led to similar reforms at the state level. As Roosevelt explained, when the spoils system encouraged corruption, "other forms of corruption will inevitably follow in its train. When a department at Washington, or at a state capitol, or in the city hall in some big town is thronged with place-hunters and office-mongers who seek and dispense patronage from considerations of personal and party greed, the tone of public life is necessarily so lowered that the bribe-taker and the bribe-giver, the blackmailer and the corruptionist, find their places ready prepared for them."[11] By the time Roosevelt was president, more federal positions were no longer part of the spoils system, with similar improvements at the state level. By 1920, half of all government jobs were filled through competitive examinations.[12] In 1903 a spokesman for New York's Tammany Hall complained that the civil service law "is the curse of the nation. There can't be no real patriotism while it lasts. How are you goin' to interest our young men in their country if you have no offices to give them when they work for their party? Just look at things in this city today. There are ten thousand good offices, but we can't get at more than a few hundred of them."[13] Urban reformers demanded competent civil service employees, including architects, engineers, and accountants to supervise improvements in infrastructure.

Businesses were also reformed. Upton Sinclair's novel *The Jungle* exposed the meat packing industry's unsavory practices, prompting laws against the adulteration of food and drugs.[14] Laws against child labor gradually removed boys and girls from mines and mills, while other laws placed them in public schools. New laws to make workplaces safer and provide worker compensation were adopted in forty-two states by 1920.[15] Finally, one of the most important changes was the introduction of a federal income tax. In 1894 Congress passed a flat tax of 2 percent on personal incomes greater than $4,000 a year, but this law was struck down in the Supreme Court by a vote of 5 to 4. To establish an income tax, the Sixteenth Amendment to the Constitution was passed by Congress in 1909 and ratified by three-fourths of the states by 1913. Only then could the federal government levy an income tax, initially for only 1 percent.[16]

The reforms described so far exemplify rational planning and civic improvements of the sort advocated by "the city beautiful" movement that sought to create attractive public spaces that could be safely enjoyed by all citizens.[17] But there was another side to the era devoted to creating social barriers and boundaries, including racial segregation, immigration quotas, Prohibition, and Americanization programs. These reforms were intended either to exclude a group or improve those who were inside the boundaries of acceptability. Some state legislatures outlawed interracial marriage. Courts upheld new segregation laws. Real estate agents, banks, zoning boards, and local authorities preserved some areas as whites-only enclaves. On another level, organizations arose to legitimate "old stock" Americans such as the Mayflower Society, founded in New York City in 1894. Membership was open only to those descended from the 102 Pilgrims who landed at Plymouth in 1620. Such genealogical societies became popular, with the greatest prestige accruing to those with colonial ancestors. The Daughters of the American Revolution (DAR), founded in 1890, excluded anyone whose ancestors arrived in the United States after 1783. In theory, it might have welcomed descendants of the 5,000 Black men who fought in the Revolution. In practice, however, the DAR was an all-white organization. In 1939, it infamously refused to let the celebrated African American soprano Marian Anderson sing in their Washington Hall. This decision so incensed Eleanor Roosevelt that she resigned from the DAR. The Mayflower Society and the DAR were gatekeepers who defined the "real Americans."

During the same years, men's clubs, women's organizations, exclusive country clubs, and golf courses were established. Most were for whites only, at times excluding Jews as well as Blacks. The first country club was established in Brookline, a wealthy suburb of Boston. Club membership became important signs of status, notably in the Knickerbocker Club of New York, the eponymous Philadelphia Club, or the Union Club and Somerset Club in Boston. The growing demand for such organizations led to the establishment of many new ones, such as Boston's St. Botolph Club (1880) and Algonquin Club (1888). Aside from proliferating clubs and genealogical organizations, membership in the elite was established by the Social Register of New York, first published in 1887, or by similar publications in other cities. In many ways, by 1929 society was less egalitarian than it had been in 1875.

Every ethnic group wanted full status. Until the middle of the nineteenth century, Jewish Americans with ancestors from Western Europe were integrated into white society. But as a tide of Eastern European Jews arrived after 1880, their Western European cousins began to experience exclusion. Likewise, the rising anti-immigration movement judged newcomers from Greece, Italy, Russia, and Poland to be less desirable than the British, French, Dutch, Germans, or Scandinavians. In 1892, the *Atlantic Monthly* published a poem by its editor, Thomas Bailey Aldrich, who described the United States as "a larger Eden planted in the wilds" threatened by "a wild motley throng" of immigrants:

Men from the Volga and the Tartar steppes,
Featureless figures of the Hoang-Ho,
Malayan, Scythian, Teuton, Kelt, and Slav,
Flying the Old World's poverty and scorn.
These bringing with them unknown gods and rites,—
Those, tiger passions, here to stretch their claws.
In street and alley what strange tongues are loud,
Accents of menace alien to our air,
Voices that once the Tower of Babel knew!

Aldrich asked, "O Liberty, white Goddess! Is it well to leave the gates unguarded?"[18]

The great energy transition was used to expand mining and manufacturing, and it also lowered the cost of transporting immigrants. Steamships were larger, safer, and faster than sailing ships, and after 1880 a one-way steerage fare could be as low as £10. The western railroads, eager to sell and populate the land along their routes, advertised and recruited new settlers in Europe. Coal mines in the Rocky Mountains recruited Mexicans.[19] Millions of first- and second-generation immigrants worked in the new high-energy industries. In 1920 they held more than half of all jobs in metal mines, quarries, textile mills, steel mills, automobile plants, shoe factories, railroads, and many other industries.[20] "Old stock" white Americans feared they would be swamped, as argued in a popular book by Madison Grant, *The Passing of the Great Race* (1916). Grant was a Yale-educated lawyer, big game hunter, champion of national parks, and supporter of the eugenics movement. He decried "the maudlin sentimentalism that has made America 'an asylum for the oppressed,' . . . sweeping the nation toward a racial abyss. If the Melting Pot is allowed to boil without control and we continue to follow our national motto and deliberately blind ourselves to 'all distinctions of race, creed or color,' the type of native American of Colonial descent will become as extinct as the Athenian of the age of Pericles."[21] Drawing on current physical anthropology, Grant ranked the intellectual capacities of racial groups. At the apex stood the Nordic race, which in his view included ancient Greece and Republican Rome. Those societies declined due to "race mixing" with the inferior populations they conquered. The United States was in danger of a similar fate. F. Scott Fitzgerald noted the popularity of such fears in *The Great Gatsby*. In the first chapter, Tom Buchanan, like Grant a Yale graduate, declares: "The idea is if we don't look out the white race will be—will be utterly submerged. It's all scientific stuff; it's been proved." He continues, "It's up to us, who are the dominant race, to watch out or these other races will have control of things." The narrator Nick Carraway, commented, "There was something pathetic in his concentration," and later concluded, "Something was making him nibble at the edge of stale ideas."[22]

Yet these "stale ideas" were popular, and legislation curtailed open immigration. The Chinese were the first to be excluded, in 1882, amid rising demands to bar other groups as well.[23] The government later denied citizenship to resident Chinese, even to those born in the United States. The Japanese were likewise perceived as part of a "yellow peril," an enormous wave of Asians threatening to overwhelm the western United States. Californian newspapers printed inflammatory stories that depicted the Japanese as degenerates and criminals. The clamor on the Pacific Coast for exclusion led President Roosevelt to negotiate an agreement with the Japanese government in 1908. Tokyo agreed to restrict passports for Japanese wives and children of immigrants who already were present in the United States. In exchange, Roosevelt ensured that new laws that excluded Japanese children from California schools would be rescinded.

White immigrants developed stories and traditions to explain why they were well-suited to become Americans.[24] Northern Europeans could point to ancestors who arrived during the colonial era, notably the French in Louisiana, the Germans in Pennsylvania, the Dutch in New York and the Hudson Valley, and the Swedes around Chesapeake Bay. They counted ancestors who fought in the American Revolution and the Civil War. Those who arrived later asserted other claims. The Norwegians argued that the Vikings had discovered North America long before 1492, as recounted in the sagas. Archaeological work in Newfoundland later proved this to be true. Before then, however, one immigrant in the Midwest carved a rune stone, buried it, and later "discovered" it as proof that the Vikings had explored the region. The Italians touted Columbus and recalled countrymen who had fought in the American Revolution, the Civil War, and the Spanish-American War of 1898. Ethnic groups erected statues to iconic American ancestors, marched in parades wearing costumes from their homeland, and succeeded in establishing special days for themselves at the world's fairs. For example, Polish Day was celebrated on October 7, 1893, at the Columbian Exposition.[25] As Orm Øverland emphasized, these practices allowed ethnic groups to affirm both their origins and their Americanness. Many immigrant communities, notably the Irish, Poles, Norwegians, Czechs, Ukrainians, and Finns, affirmed allegiance to the United States and celebrated American heroes of their ethnicity. At the same time, they advocated independence and democracy for their homelands. Immigrant communities felt that multiple loyalties were united in one democratic identity; they did not need to choose.

Recent immigrants created ingenious arguments to demonstrate a special affinity between their homeland and the United States. The Greeks argued that they had invented democracy and ideas embodied in the American Constitution. Their ancient architecture provided the model for courthouses and public buildings across the nation. Their literature was canonized as the foundation of Western culture, and it was taught at every American university. Furthermore, an ancient Greek had discovered that the earth is round, and this idea was a necessary precondition for Columbus to sail west and discover the New World. Therefore, Greek immigrants naturally belonged in the United States![26]

Every immigrant group claimed that they brought special qualities to America. These arguments took the form of the "cultural gifts movement" during the 1920s, a period when immigration was restricted, with quotas that favored Northern Europeans over Southern and Eastern Europeans. The cultural gifts movement rejected the assimilationist assumptions of "the melting pot." It built on the ideas of John Dewey, Jane Addams, and others who before World War I had "argued for the preservation of ethnic cultures."[27] They focused on education as the means to conquer prejudice and preserve a colorful American diversity. The movement would gain strength in the 1930s, when the triumph of fascism in Europe increased the need to combat prejudice. To some extent the idea of cultural gifts relied on stereotypes (such as musical Italians or stoic Chinese) and it urged preservation of some admirable qualities and characteristics, such as ethnic cuisines, but not preservation of an entire heritage. The goal was a picturesque variety, not alternative cultures.

In contrast, homosexuals remained in the shadows, as their sexual preferences were defined as either a perversion or a form of mental illness. Sodomy was a crime in every state of the nation, though enforcement varied. As George Chauncey explains, a gay subculture did exist in New York and other large cities after 1880, and it was not entirely hidden.[28] Unlike the rather strict dichotomy between heterosexual and homosexual that became the norm by World War II, Chauncey argues, before then sexual identity was not defined by biology nor was it fixed, as some married middle-class men had recreational gay sex, and some occupational groups such as sailors and police officers were likely to include bisexuals. Between 1875 and 1930 such sexual activity was not acknowledged in polite society or represented in the cinema and popular culture. Men could not be openly homosexual, for that was defined as a crime. Indeed, the word "homosexual" was not even used by the *New York Times* until 1926, when it appeared in a discussion of the young Chicago murderers Nathan Leopold and Richard Loeb, who kidnapped and killed a fourteen-year-old boy.[29]

Women fought for full acceptance as citizens, demanding not only the right to vote but also the opportunity to become doctors, lawyers, architects, or university professors. They finally began to reach some of their goals after the Civil War. In 1869, the first female lawyer in the United States was admitted to the bar in Iowa, and a decade later a law permitted women attorneys to argue cases before all federal courts. Yet the first women were not admitted to the American Bar Association until 1918, and the first woman was not appointed to be a federal court judge until 1928. At the same time, more women were taking jobs in factories, department stores, telephone exchanges, and offices, where they operated new technologies such as typewriters, mimeographs, cash registers, and tabulators. Such work required not so much physical strength as precision and accuracy.

Increasing employment for women occurred in tandem with the national suffrage campaign and demands for equal pay. Its leaders were primarily "old stock" Americans, in some cases with a Quaker background, notably Susan B. Anthony, Alice Paul, Elizabeth Cady Stanton, Lucy Stone, and Ida B. Wells. Many had also supported the abolition of slavery, and after the Civil War the movement was split for a generation between those who linked women's suffrage to Black suffrage and those who treated the two as separate issues. Success for white women came when the two issues were separated. White women won the franchise state by state first in the West and then in the Midwest and Northeast. Marches and demonstrations grew larger and by 1913 would draw thousands of supporters to march in Washington, DC, the day before Wilson was inaugurated (figure 5.1). Women got the vote in time for the 1920 elections. However, the struggle for equal pay continues.

Figure 5.1
Women's suffrage rally, Washington, DC, 1913. G. V. Buck, photographer. Courtesy of the Library of Congress, Prints and Photographs Division.

While white women and European immigrants could claim full membership in society, this proved impossible for nonwhites. A number of supposedly scientific theories of racial differences defined them as inherently unequal. Some nineteenth-century scientists like Harvard University's Louis Agassiz believed in polygenism, the idea that the different races were distinct species that had evolved separately. Attempts to prove this theory focused on the interpretation of fossils and pseudoscientific methods like craniometry (the measurement of human skulls). The overriding conviction was that white people were biologically superior. Such ideas were adopted at the Smithsonian Institution and other scientific organizations, which embraced the idea of a hierarchy of the races. Museum exhibits presented this hierarchy as a given, and this apparent fact also justified the emerging eugenics movement.[30]

Some climate scientists advanced another theory that seemed to explain white superiority. In 1915 Yale University Press published Ellsworth Huntington's *Climate and Civilization*. As Rebecca Wright notes, it argued that "racial difference was a result of climatic variation and that the superiority of the white Anglo-Saxon race (over other racial groups) could be attributed to the potency of climatic energy that had driven the evolutionary process forward" in temperate regions. The climate literally produced American exceptionalism, because the natural distribution of energy placed "white Anglo-Saxon bodies at the top of the racial hierarchy." The differences were stark: "The white body was characterized by high levels of energy, vigor, and efficiency. In contrast, the black body was defined by torpor, friction, and lack of momentum."[31]

Theories of "social Darwinism" also flourished in these years. They claimed that those who attained power and wealth did so because there were innately stronger and more intelligent than the weak and impoverished lower classes. From this perspective, the power and wealth of white Americans demonstrated their superiority and legitimized their privileged position. Similar thinking suggested to Andrew Carnegie and other industrialists that the rise of large corporations and the demise of smaller companies exemplified "the survival of the fittest." These ideas not only were accepted as science but also shaped exhibits at the major world's fairs and their ethnological villages. The historian of expositions Robert Rydell notes that "the non-white people living in villages along the Midway not only were seen through the lens of America's material and presumed racial progress leading to a future utopia but were neatly categorized into the niches of a racial hierarchy."[32] At the Columbian Exposition, the German and Irish villages were located close to the educational exhibits, while the Muslim world was placed in the middle of the entertainment zone. Further off, at the edge of the fairgrounds, were the primitive villages of Africans and Native Americans. As the *Chicago Tribune* summarized this aspect of the Exposition, the visitor could "descend the spiral of evolution, tracing humanity in its highest phases down almost to its animalistic origins."[33] The Buffalo Exposition of 1901 provided a similar display of ethnological villages, inhabited by "genuine primitives." The exposition's lighting featured harsh, crude colors around these villages, giving way to more aesthetically pleasing color schemes for the educational exhibits at the center of the fairgrounds. As Rydell explains, that exposition layout was a "carefully crafted allegory of America's rise to the apex of civilization."[34] The idealized lifeworld constructed at the world's fairs was a hierarchical society where the white race was naturally superior.

In harmony with these theories and exposition displays, Reverend Josiah Strong's bestseller *Our Country* argued that the Anglo-Saxon race was more vigorous and intelligent than the other races and was destined to dominate the world. He argued that "the aborigines of North America, Australia and New Zealand are now disappearing before the all-conquering Anglo-Saxons. It seems as if these inferior tribes were only precursors of a superior race" that would overwhelm them on their way to domination.[35] Racism based on supposed science could also be found in mass distribution magazines such as the *Saturday Evening Post*, whose editor believed that Southern European immigrants threatened the nation. He declared in 1919 that without restrictions, "the melting pot will end up as a garbage pail."[36] His magazine had a circulation of two million. During World War I, Americans began to fear German immigrants. "Anglo-Saxon Americans began casting themselves as soldiers of the Almighty," and President Wilson declared that the war against Germany was a crusade to make the world "safe for democracy." Wilson created by executive order a Committee on Public Information to convince the country to support the war, using posters, cartoons, pamphlets, and films. Conscientious objectors to the war who refused to wear uniforms and perform hospital service were imprisoned and treated harshly.[37] Fear of contamination by immigrants and their dangerous ideologies reached a frenzy after the Russian Revolution, when these fears were combined with a dread of anarchism and communism. Newspaper cartoonists "depicted the Bolshevik and German threats as indistinguishable."[38]

Immediately after the war, the 1919 "Red Scare" rocked the nation, under the slogan "one hundred percent Americanism." In January 1920, the Justice Department, working with local police, staged simultaneous raids across the nation. They arrested more than 3,000 immigrants who were alleged socialists, communists, or anarchists. Many were from Ukraine, Lithuania, Poland, Russia, and other parts of eastern Europe. Attorney General A. Mitchel Palmer defended this dragnet, which he said had dredged up "alien filth" and "alien criminals." He declared in *The Forum* magazine, "Like a prairie-fire, the blaze of revolution was sweeping over every American institution of law and order a year ago. It was eating its way into the homes of the American workman, its sharp tongues of revolutionary heat were licking into the altars of the churches, leaping into the belfry of the school bell, crawling into the sacred corners of American homes, seeking to replace marriage vows with libertine laws, burning up the foundations of society."[39]

The Red Scare was an extreme expression of a more general determination to protect and purify the nation from the perceived threats of inferior races and deviant ideas. These efforts focused on radicals but even more on Black Americans. It was not enough merely to declare white superiority. Highly visible actions drew the boundaries. Many amusement parks prohibited African American visitors, and the great world's fairs often excluded them as well, except for certain "Negro Days" when they could attend. The former Confederate states legalized systematic racial segregation in all aspects of life. Ninety percent of the Black population then lived in the South, and they had to use separate schools, churches, railway carriages, entrances to public buildings, and even separate African American cemeteries.

There was less comprehensive segregation in the North, where increasing numbers of Blacks migrated, attracted by opportunities in the high-energy industrial economy. When World War I both curtailed immigration and drew many men into the army, jobs opened up in Chicago, Detroit, Baltimore, Philadelphia, and New York, attracting thousands of Black Americans. After the war, these migrants competed for jobs with white veterans. Labor friction intensified with the return of thousands of Black veterans who had served in France, where they had experienced less racial discrimination than at home. The situation was explosive. The economy was temporarily weak and unemployment high, as many industries halted war production and closed to retool.

In 1919 more than twenty riots by whites against Blacks occurred across the United States. David Goldberg summarizes: "In 1919, the mere presence of an African American soldier could provoke a violent response. In Chattanooga, Tennessee, crowds tried to stop black veterans from wearing their uniforms. In the small southwestern Georgia town of Blakely, a mob killed an African American soldier. In Vicksburg and other Mississippi communities, attacks on African American soldiers became common during 1919."[40] In Charleston, South Carolina, a white riot killed three Black men. In the early summer came more violence in Texas, Arizona, and Indiana. But the worst riots occurred during the sweltering days of July. In Washington, DC, white veterans, some in uniform, attacked the city's Black neighborhoods. James Weldon Johnson declared, "I knew it to be true, but it was almost an impossibility for me to realize as a truth that men and women of my race were being mobbed, chased, dragged from street cars, beaten and killed within the shadow of the dome of the Capitol, at the very front door of the White House."[41] After two days of violence, President Woodrow Wilson called in 2,000 troops to reestablish order. No sooner was that mob quelled than another battle occurred in Norfolk, Virginia. There, returning African American veterans held a homecoming celebration that a white mob attacked. Six suffered bullet wounds and two died.

The worst riots occurred the following week in Chicago.[42] They began on the shores of Lake Michigan when some African American bathers crossed an informal barrier and entered a "white beach." Whites pelted them with stones, killing one and sparking a riot that lasted for days. White mobs, with only token opposition from the police, attacked Black neighborhoods. Many of the aggressors were from Irish neighborhoods adjacent to the Black ghetto on the South Side, whose population had doubled since 1910. The unrest lasted a week, and at least thirty-eight people died, and more than five hundred were severely injured. The long-term effect of these attacks was both to concentrate Blacks in ghettos and to increase resistance to racism through organizations such as the NAACP.

The riots of 1919 were not an aberration, but part of a continual white insistence on hard racial boundaries that were being undermined by internal migration, urbanization, and a changing job market. In 1921 whites attacked the local African American population in Tulsa, Oklahoma.[43] Its 10,000 Black residents lived in the Greenwood neighborhood, including a prosperous business district. Anger boiled over in the white community amid rumors (later proved false) that a black man had sexually assaulted a white woman in an elevator. On May 31, a mob assembled at the jail and demanded that the young man be handed over so they could lynch him. The police refused. The white mob then assaulted the Greenwood area, burning houses, businesses, a library, churches, and even a hospital. They murdered many (the precise number remains unknown) and they drove away firefighters who attempted to minimize the destruction. In all, thirty-five blocks were destroyed, and 8,000 people were rendered homeless. In the aftermath, 6,000 African American citizens were herded into a local fairground and "protected" there by armed guards. In both Chicago and Tulsa, whites exaggerated the situation. The Tulsa mob imagined they were preventing an insurrection.

In addition to Jim Crow laws and race riots, whites made deadly attacks on individual Black people. The National Association for the Advancement of Colored People (NAACP) published a report in 1919 that detailed 3,224 lynchings that had occurred since 1889. This epidemic of white violence occurred at a rate of more than one hundred lynchings each year.

Figure 5.2
Ku Klux Klan parade, Pennsylvania Avenue, Washington, DC, September 13, 1926. Aerial photo, National Photo Co. Courtesy of the Library of Congress, Prints and Photographs Division.

It continued after World War I. In 1919, eighty-three were lynched, in 1920 sixty-four, and in 1922 sixty-one.[44] During these years the Ku Klux Klan revived and recruited thousands of members. Membership was limited to native-born citizens. When joining, recruits had to pledge that they wanted limits on "foreign immigration" and that they believed "my rights in this country are superior to those of foreigners."[45] The Klan was powerful in the South, but it also had many members in northern states, such as Ohio, Kansas, or Indiana, where 100,000 people once attended a rally in Kokomo. At its height in 1924, the Klan had 4.5 million members, and staged marches in major cities, including Washington, DC (figure 5.2). In

Figure 5.3
Children assembled at New York's City Hall for an Americanization event, September 27, 1917, Bain News Service. Courtesy of the Library of Congress, Prints and Photographs Division.

the South, the perceived threats came from Blacks. In the North, the Klan railed against Jews, Catholics, and immigrants. It warned members against the Pope, whom they feared would take over the United States. The Klan was strong among "old stock" Protestants, notably in Maine, Michigan, Oklahoma, and Colorado. But the Klan was not as strong among farmers as in small cities and county seats, and it did not appeal to many factory workers. It had only a tiny membership in Boston, Philadelphia, or Cleveland. Although the Klan declined in size and influence, many politicians retained its anti-immigrant rhetoric. After World War I, immigration laws were redefined, creating a quota system that favored Northern Europeans over more recent immigrants from Southern and Eastern Europe. The ideal immigrant was defined as a heterosexual white Protestant. The undesirable "others" were gay, Black, Asian, Southern European, Muslim, or Hispanic.

Many recent immigrants were thought to be poorly integrated into society, and before World War I an intensive effort began to "Americanize" them. Schools and local governments took part, including mass meetings and rallies (figure 5.3). Corporations offered classes inside factories to prepare foreign workers for citizenship. The classes were often taught by junior executives. Instruction aimed to improve fluency in English as well as to provide an understanding of democratic institutions. Americanization was part of a larger effort to combat radicalism. An Inter-Racial Council established in 1919 sought "To stabilize industrial conditions. To apply American business methods to the foreign-born press by building up an American advertising base under it. . . . To decrease radicalism" and to counter "the foreign language press dealing with attacks upon American institutions, law and order and industry."[46] Educating the foreign-born worker was expected to undermine radicalism and promote the idea that United States was a cultural melting pot. At the Ford factories, ceremonies were held to celebrate workers who had acquired citizenship. Each would appear on stage dressed in the clothing of their homeland. They would then step behind a large painting of a melting pot and emerge a few moments later dressed in a suit and waving an American flag. They had abandoned their foreign identity and entered the charmed circle of the kaleidoscope.

A racial group was either in or out. Unlike Blacks, Native Americans were regarded as potentially white. Their children were often removed from families, sent to boarding schools where their language and customs were banned, and brought up as though they were white. In contrast, Asians, Blacks, and Hispanics could not become white, and in many states, they could not marry a white person either. Discrimination against those considered undesirable was normalized. This was particularly clear in the segregated South. During the great flood of 1927, when the Mississippi River broke through levees and threatened the lives and property of all citizens, inequality in rescue operations was taken for granted. The poet Langston Hughes visited Louisiana soon afterward and learned that "the white refugees were brought down the river to the city in steamers with cabins and covered decks to protect them from the elements, while the Negroes were transported on open flatboats, exposed to the wind and weather. In Baton Rouge, the Red Cross had housed the whites in a group of tree-shaded buildings that were former government barracks, I believe. The Negroes were housed in an open field in small tents, where the mud was ankle deep when it rained. The whites were given three hot meals a day, the Negroes, two. The whites received regular rations of tobacco, snuff, and candy. The Negroes got what was left over, if any, of these delicacies."[47] This overt discrimination in the assistance to flood refugees was paid for by the Red

Cross but carried out by its local chapters and local governments. In some cases, African Americans were housed in camps along the levees and prevented from leaving by the National Guard, while white planters were put in charge of distributing food and supplies.[48]

Such discrimination was justified by "scientific" racism. In 1929, during a public debate with W. E. B. Du Bois, Theodore Lothrop Stoddard, a Harvard PhD and prominent author, declared: "We know that our America is a White America." He believed that "the overwhelming weight of both historical and scientific evidence shows that only so long as the American people remain white will its institutions, ideals, and culture continue to fit the temperament of its inhabitants and hence continue to endure."[49] Maintaining racial purity, in this view, was an essential bulwark against national decline.

If one believed such things, then it was also easy to believe that antebellum plantation slavery had been a beneficent and even educational system. This argument minimized the importance of abolitionists and slave resistance, and it absolved slaveholders of being unjust. They were only doing what was necessary at that earlier historical stage. As Jennifer Lieberman notes, the electrical industry adopted this argument, in advertising that referred to new appliances as "electric slaves." A 1915 General Electric advertisement, "Slaves of Yesterday and Today," juxtaposed building the Egyptian pyramids using 100,000 enslaved people with how a "modern skyscraper goes up in a fraction of a year—workmen furnishing the skill and General Electric motors the muscle." In this imaginary, not reformers but machines had abolished slavery, which otherwise might still be necessary. The *General Electric Review* made this connection explicit: "the old successful democracies" all "had slaves to cultivate their fields, dig their ditches, wash their plates and dishes, and to do all those things that no one wants to do. These slaves had no part in the democratic life of the state." In Athens and other early democratic societies, "it was not until they had these slaves that they could devote their time to that literature and art which has made then famous."[50] Similar ideas were repeated by a team of lecturers, using a slide presentation created by Charles M. Ripley for the National Electric Light Association. In 1927, audiences for this talk included businessmen, civic organizations, and university engineering students, who learned that new machines produced equality as they liberated humanity from toil.

Faced with deterministic theories of history and scientific theories of racism, African Americans still found ways to advance. The most prominent example was Booker T. Washington. Born enslaved in 1856, after the Civil War he received an education as a skilled worker at the Hampton Institute. When just twenty-five he became the first leader of the new Tuskegee Institute in Atlanta. His first students learned to make bricks and acquired construction skills, so that they could put up the school's buildings. Tuskegee eventually offered a practical education in thirty-eight trades and professions, including teaching, farming, carpentry, cabinetmaking, printing, and shoemaking. Some courses were for women only, for example, instruction in domestic skills such as sewing and cooking. By the time of Washington's death in 1915 Tuskegee had a faculty of two hundred, more than a hundred buildings on a 2,000-acre site, and an endowment of $2 million. His success depended on his skills as a speaker and fundraiser.[51] He gained the trust of prominent white philanthropists and received contributions from three of the wealthiest men in the United States: J. P. Morgan, John D. Rockefeller, and Andrew Carnegie. Washington promoted the idea that Black Americans could lift themselves up through practical education in agriculture and trades. He argued that skilled black workers were essential to the growth and success of the South. As he put it in a famous address to a white audience at the Atlanta Exposition, "we have proved our loyalty to you in the past, in nursing your children, watching by the sickbed of your mothers and fathers, and often following them with tear-dimmed eyes to their graves, so in the future, in our humble way, we shall stand by you with a devotion that no foreigner can approach, ready to lay down our lives, if need be, in defense of yours, interlacing our industrial, commercial, civil, and religious life with yours in a way that shall make the interest of both races one. In all things that are purely social we can be as separate as the fingers, yet one as the hand in all things essential to mutual progress."[52] Using such arguments, Washington was able to expand and endow the Tuskegee Institute. He became a national figure, and he was the first Black man to dine at the White House, invited by Theodore Roosevelt. Outraged southern newspapers editorialized against it, and no other Black leader dined there for a generation.

In contrast to Washington's cooperative boundary maintenance, younger Black leaders challenged racial barriers and demanded an end to segregation. The most notable figure was Du Bois, who called a meeting of those ready to fight for equality, held in Niagara, Canada, in 1905. Those who attended rejected the idea that "the Negro American assents to inferiority, is submissive under oppression and apologetic before insults." Their resistance led to the establishment of the National Association for the Advancement of Colored People in 1910. The NAACP in its early years counted many whites as members, and it preferred to work through the courts rather than agitate in the streets. It gradually established legal precedents and protections that increased Black equality, in a process that took half a century. The NAACP drew on the talents of Black leaders such as Charles Hamilton Houston. The valedictorian of the class of 1915 at Amherst College, he served as an officer for a racially segregated artillery unit in France during World War I. Houston graduated from Harvard Law School in 1923 and was the first Black student elected to the law review. He later created a law curriculum for Howard University and joined the NAACP as first counsel.[53] Throughout his life, he fought for equality through the courts. Houston embodied Du Bois's idea that a "talented tenth" could open the doors of opportunity for other Black Americans.

Du Bois understood the psychological cost of racism. A famous passage in *The Souls of Black Folk* deals with "double consciousness, this sense of always looking at oneself through the eyes of others, of measuring one's soul by the tape of a world that looks on in amused contempt and pity. One ever feels his two-ness,—an American, a Negro; two souls, two thoughts, two unreconciled strivings; two warring ideals in one dark body, whose dogged strength alone keeps it from being torn asunder."[54] Double consciousness was a cognitive dissonance imposed on Black people by white society. Double consciousness was an awareness of the artificial and partial nature of kaleidoscopic vision, which relied on one eye only, producing a flattened, simplified, and homogenized view of American life. As Du Bois elaborated in a later chapter:

> From the double life every American Negro must live, as a Negro and as an American, as swept on by the current of the nineteenth while yet struggling in the eddies of the fifteenth century,—from this must arise a painful self-consciousness, an almost morbid sense of personality, and a moral hesitancy which is fatal to self-confidence. The worlds within and without the Veil of Color are changing, and changing rapidly, but not at the same rate, not in the same way; and this must produce a peculiar wrenching of the soul, a peculiar sense of doubt and bewilderment. Such a double life, with double thoughts, double duties, and double social classes, must give rise to double words and double ideals, and tempt the mind to pretense or revolt, to hypocrisy or radicalism.[55]

Even as Du Bois wrote these words and the NAACP waged its legal battles, some Black athletes demonstrated their equality, and occasionally their superiority. A notable example was the bicycle racer Marshall "Major" Taylor, the "Worcester Whirlwind."[56] In 1892, at age fourteen he was hired in Indianapolis as a stunt man to entertain the crowds between bicycle races. Already the following year he was one of the racers. He soon moved to a team in Worcester, Massachusetts, where racial prejudice was not as strong. But as Taylor defeated white men, some competitors resented it. When he went south to train during the winter, as other professional riders did, he was excluded from a track in Florida, and he was forced to flee Savannah, Georgia, after receiving a threatening letter: "If you do not leave here before 48 hours, you will be sorry. We mean business. Clear out if you value your life."[57] His New York promoters arranged alternate training on Long Island. In 1898 he set a world record for the mile and won many races. When just twenty-one, he was the first Black American in any sport to become world champion, winning that title in Montreal. He then traveled to Europe, where he won many races.

Another athlete born in 1878, the boxer Jack Johnson, also challenged the color line. Despite obvious talent, he had difficulty getting a match against the best white American fighters, notably the heavyweight champion, James Jeffries. Johnson had to go abroad in 1908 to become world champion, by defeating Tommy Burns in Australia. Then Jeffries came out of retirement, "for the purpose of showing that a white man is better than a Negro." They fought in Reno on July 4, 1910, in front of an all-white crowd of 22,000. Before the match, a band played "All Coons Look Alike to Me." Johnson knocked Jeffries down twice and won the fight in the fifteenth round, followed by racial disturbances in fifty cities across the United States. Johnson defied segregationists by marrying white women, and in 1911 he found it necessary to flee to France after his second such marriage. Other sports also had talented Black stars, notably in the segregated Negro baseball leagues. The all-white Major Leagues resisted integration until 1947. Before then, occasional interracial games showcased some of the great Black players.[58] In cycling, boxing, baseball, and many other sports, Black athletes proved they were equal.

These demonstrations of prowess angered many whites, violating their belief that the white race was more highly evolved, as the Smithsonian exhibits at world's fairs apparently demonstrated. Blacks, they thought, belonged to a lower rank of humanity. This meant that "miscegenation" or "racial mixing" was akin to pollution. An elaborate, specialized vocabulary was developed to specify the degree of blackness for any offspring of such unions. Some of these terms were "mulatto" (half Black), "quadroon" (one-quarter Black), "octoroon" (one-eighth Black), "sextaroon" (one-sixteenth Black), and "sagmelee" (one sixty-fourth Black). A person with even "one drop" of Negro blood was regarded as Black. More than half of the states outlawed miscegenation, or marriage between Blacks and whites, which the Supreme Court did not declare unconstitutional until 1967. Such marriages were criminalized throughout the South, but they were legal in every state in the North and much of the Midwest. This was a rural/urban divide as well, for interracial marriages were more likely to be tolerated in cities than in the countryside.

To escape discrimination, thousands of those with light skin slipped over the color line into white society. When successful, "passing" offered immediate advantages in employment and freedom of movement, but it meant cutting family ties and denying one's heritage. As Allyson Hobbs notes, this "chosen exile" became common as Jim Crow laws were passed, and African Americans migrated into northern and western cities. Those who passed into white society did so at the price of a constricted identity.[59] Du Bois had described the double consciousness of Black people confronting racism, but anyone "passing" had to develop "triple consciousness," adding another layer of complexity. They had to pretend to be white and outwardly seem to accept racial inequality. They could not risk open opposition to a racist joke or discriminatory behavior toward a Black person. They had to seem unaffected by segregated transportation or separate entrances for each race. That triple consciousness arguably entailed an even greater "wrenching of the soul" than double consciousness. Anyone "passing" had to lead a triple life and perpetrate the very racism they abhorred and sought to escape.

Double consciousness was a problem not only for Blacks but also for whites, many of whom did not recognize their schizophrenic social behavior. Many white Americans lived by a double standard. They behaved one way toward other whites and quite a different way toward Black people, whom they commonly addressed as unintelligent inferiors who owed them deference. Black people were expected to get out of the way on the sidewalk and to act the part of a servant. Black men, regardless of age, were addressed as "boy." Whites believed in a racial hierarchy that bolstered their sense of entitlement. They felt insulted should an "uppity" Black person fail to perform as a friendly inferior. In this construction of social relations, whites were haughty, easily angered, and righteously violent in defense of their presumed superiority. They were not ready for an open and honest dialogue, nor could they understand that racism compromised their own humanity. It was a time of latently violent social control that could culminate in lynching and race riots. As Walter Lippmann observed after the Chicago riots of 1919, "those who degrade and terrorize are inevitably themselves degraded and terror-stricken."[60]

Racial divisions were never absolute, however. In popular culture whites and Blacks learned from one another, notably when performing ragtime, blues, and jazz. Whites went to Harlem to hear music at the Cotton Club, where the audience was often white, the musicians Black. Racial lines could fade away in recording studios, jam sessions, and eventually in some concerts. In the workplace, too, as Black people became clergymen, doctors, dentists, lawyers, and accountants, they quietly challenged segregation in their professions. They were at times supported by whites opposed to discrimination, and progress toward racial equality inched forward.

Imposing boundaries was not a simple matter. There were always gray areas with unclear dividing lines. There were always some who could successfully pretend to be white. Some activists refused to accept the legitimacy of the artificial boundaries, and some refused to abide by the boundaries, like the boxer Jack Johnson. Yet these artificial divisions had psychological consequences, forcing Black Americans and other persecuted groups to adopt a watchful "double consciousness," to defend themselves against white Americans and the double standard. The boundaries that defined inequality were internalized as psychological barriers that increased racial tensions and created misunderstandings.

While racial exclusion and discrimination were dominant practices between the end of Reconstruction and 1929, white America was not a monolithic privileged group inside a charmed circle. White America itself was riven with tensions between management and labor, city and countryside, immigrants and nativists, and new technologies and old practices. Whites had a privileged position compared with other racial groups, and white men were the most privileged of all, but nevertheless their lives were marked by labor conflicts, disrupted by new machines and processes, and convulsed by rapid growth. White people were the backbone of most reform movements, whether to create public parks or end corrupt politics.

But some "reforms" were intended to change people. In the 1890s the "Pledge of Allegiance" to the national flag was invented and promulgated as the proper beginning of every school day.[61] Some reformers wanted to Americanize immigrants. Others sought to abolish the "white slavery" of prostitution. Their rhetoric emphasized sin and salvation, often without reference to the poverty that might push women into sex work.[62] When reformers campaigned against red-light districts and closed brothels, they drove sex workers into the arms of pimps in the streets. A less moralizing approach focused on protection from sexually transmitted diseases that could spread back home to spouses. As medical issues became salient, the moralism of the American Purity Alliance gave way to the regulatory focus of the American Social Hygiene Association. It viewed sex work in terms of women's health, infant mortality, birth defects, and mental disabilities. However, by the 1920s, it was clear that neither moralistic laws nor medical intervention could eliminate the "world's oldest profession."

Another purity crusade campaigned to pass the Eighteenth Amendment to the Constitution, which prohibited sale of beverages that contained more than 0.5 percent alcohol. Prohibition was supported by the Anti-Saloon League and the Women's Christian Temperance Union.[63] It was expected to close saloons, reduce crime, eliminate drunk driving, empty prisons and poorhouses, enhance the productivity of workers, and improve public health. Even more than the crusade against prostitution, however, Prohibition failed. Saloons did close, but speakeasies replaced them. Criminals built distilleries that made bootleg liquor, and they imported wine from Europe, whiskey from Canada, and rum from the Caribbean. The expense of enforcing Prohibition doubled between 1921 and 1930, as millions of people flouted the law. It may have reduced alcohol consumption in the first years but not in the long run. Prohibition criminalized half the population and undermined relations between citizens and the police.

Like efforts to curtail prostitution and prohibit alcohol, moral reforms often backfired, while reforms to improve infrastructure, increase democracy, or regulate business usually endured. Overall, there were five sorts of reforms. One central impulse was to improve public amenities and infrastructure, including paved roads, mass transit, parks, playgrounds, reservoirs, swimming halls, and other sports facilities The new energy systems also required installation of networks for electricity, gas, telegraph and telephone lines, water, sewage, traffic control, police call boxes, fire alarms, and fire hydrants. Another obvious area of reform was industry. New laws were passed to outlaw child labor, to minimize smoke pollution, to restrain corporate power through regulation, to eliminate monopolies, and to prevent the excessive accumulation of wealth through an income tax. A third cluster of reforms democratized the political system: women's suffrage, direct election of senators by popular vote, and the right to hold referendums. The spoils system was replaced by a civil service based on merit. Most such reforms succeeded. Children went to school instead of working in mines and factories. Oligopolies replaced monopolies. The Senate became a less exclusive club, and the civil service functioned better.

But if reforms of infrastructure, business, and politics were largely successful, both exclusionary and inclusionary reforms were not. The "reforms" that segregated races and excluded "undesirable" immigrants were not durable. They expelled certain groups from the national imaginary, but in the long run they could not be erased. Prohibition did not stop the consumption of alcohol, and it did not reduce crime but increased it. Forcibly removing Native American children from their families and sending them to boarding schools was traumatic and undemocratic, and eventually was abandoned. By comparison, inclusionary reforms did less long-term harm, as they attempted to Americanize, educate, or otherwise improve white people. Americanization programs did help immigrants become citizens. More children were in school, even if they began each day by reciting the "Pledge of Allegiance." Settlement houses provided a venue for mutual understanding. Hundreds of new libraries and museums increased access to knowledge and culture. Suffrage improved female equality and made it possible for women to have more direct influence.[64]

Reforms of exclusion and inclusion both idealized a white Protestant America that included white ethnic groups but rejected other differences. Between 1910 and 1920 the impulse to define and enforce that society's boundaries peaked just as the great energy transition reached the penultimate stage of technological momentum. The often angry determination to erect barriers was not sustained, however. Americanization faded away and the Klan declined, but the NAACP continued its long fight against discrimination. Even in the 1930s the barriers began to erode. The reforms such as paved roads, streetlights, direct election of senators, and the income tax endured, but Prohibition, Americanization, segregation, and most other "moral reforms" did not.

6

NOSTALGIA

Faced with the often-rapacious engineering of nature, the geographical dislocations caused by a major energy transition, the turbulent job market, the disintegration of time and space, the malleability of sound and image, and the enormous scale of an industrial economy, many Americans sought solace in recollections of previous lifeworlds. Their nostalgia was often illusory, as they recalled inaccurately and selectively, but these half-truths strengthened the sense of having roots and reinforced the sense of identity and community.[1] Americans were variously attracted to four idealized pasts. One was the antebellum period. This lost era was recalled in commemorations, parades, statues, and memorials, many recalling the American Revolution or the values it represented. Second, others looked to the more distant, European past. These anti-modernists valued artisanal production, gothic architecture, classical music, and other aspects of a pre-industrial life. Third, many Americans became nostalgic for the small town and rural life. Fourth, the country began to set aside state and national parks that preserved "untouched nature" for future generations. Each of these four imaginaries was a reaction against the turbulent present in favor of a former lifeworld.

The antebellum era was often memorialized in monumental structures. Building the Washington Monument had begun in 1848, but it was then delayed by political infighting, lack of funds, and the Civil War. During Reconstruction, however, the need for unifying symbols seemed greater than before. Celebrating the general of the revolutionary armies and the first president became a priority, not least because he was a Virginian who commanded the respect of both the North and the defeated Confederates. In the centennial year of 1876, Congress appropriated the necessary funds and work began again. The monument was completed in December 1884, and dedicated on February 22, 1885, Washington's birthday.[2] It was the tallest structure in the world until surpassed by the Eiffel Tower in Paris. After several speeches and President Chester A. Arthur's brief dedication, a procession of more than 6,000 people formed, drawn from all parts of the nation, including the South.[3] As they marched by, the public saw a demonstration of unity and fealty to the memory of Washington. Troops came not only from nearby Alexandria, the District of Columbia, and Maryland but from across the country, including Boston, Hartford, Detroit, Charleston, and Richmond. There were many bands to enliven the proceedings, including one from the Naval Academy. A long column of Masons from the Grand Lodges of many states marched by, proud that Washington had been a member of their order. An entire section of the parade consisted of carriages carrying state governors, members of Congress, and other dignitaries. Nor were the carpenters and stone masons who had erected the monument forgotten, as they too were honored marchers. The parade's organization in no way referred to the Civil War. It mixed troops from all parts of the country, put the governors from the North and South together, and in all its arrangements dissolved sectional differences. Everyone could agree on the greatness of Washington. The dedication and parade served to bind the nation together. The theme of unity was sounded again in an address to Congress after the dedication, which included an observation that Washington made in 1783 just after national independence had been attained. In a letter, he stated that "the well-being—I may even venture to say the existence—of the United States as an independent power" depended on "an indissoluble union of the States under one Federal head."[4] The monument, the parade, the ceremonies, and the final oration evoked the glory of Washington and deleted the fratricidal war between the states. As the South was reinscribed within the nation, the fact that Washington had owned enslaved people was never mentioned.

The following year another symbol was dedicated in New York's harbor, the Statue of Liberty. If Washington was the model of patriotism and self-sacrifice, the great statue honored liberty as a fundamental value the United States shared with France. Most Americans had heard of Patrick Henry's famous speech made in 1775 at the dawn of the American Revolution, which concluded, "Give me liberty or give me death." A year later, Thomas Jefferson wrote the famous preamble to the Declaration of Independence that began with the assertion, "We hold these truths to be self-evident: that all men are created equal; that they are endowed by their Creator with certain unalienable rights; that among these are life, liberty, and the pursuit of happiness." These words were implicitly understood to apply to white

Americans, primarily men. Equality for women, including the right to vote, had been achieved in only a few states by 1886, and Blacks and others were being marginalized at the same time as the Statue of Liberty was erected on its pedestal.

Nevertheless, in the international context of 1886, the United States could claim to be in the world's vanguard of democracy, compared with Germany, Russia, China, Japan, and many other nations. On October 28, a great parade celebrated the statue's dedication. Led by President Cleveland, it included representatives of other democracies, including France, Italy, Belgium, and Switzerland. One hundred bands braved a steady rain and marched in the vast parade. An estimated one million people watched it wend its way down Manhattan toward the tip of the island facing the statue. When the parade reached the financial district, office workers began spontaneously to throw down ticker tape, and from that moment ticker tape parades became a tradition. The president and foreign dignitaries then took a boat to Bedloe's Island for the dedication, which the crowds along the shoreline could not hear. A large French flag covered the statue's face, and it was to drop away after the speeches. By accident, the unveiling occurred prematurely, which prompted all the ships in New York Harbor to blow their whistles and let off a good deal of smoke. The crowds assumed the formalities were complete and began to cheer, and cannons fired celebratory salvos. Because of the rain, the fireworks were postponed until November 1.[5]

After its dedication, the Statue of Liberty did not instantly become a symbol of America. While a steady stream of tourists visited it, for decades it was neglected by the federal government, which skimped on maintenance and constantly had to be prodded by private associations to protect the monument and develop the site. Bedloe's Island was ill-kept and covered with ramshackle warehouses, and the dock where tourists landed was an eyesore.[6] Despite this negligence, however, the statue grew in symbolic significance. Except for Brooklyn Bridge, it was the most prominent sight in New York's harbor.

The meanings given to the statue changed, as the public began to connect it with the promise of the new world for immigrants. This was not part of its intended meaning. Frédéric Auguste Bartholdi, the sculptor who created the statue, had meant to commemorate the friendship of France and the United States and their shared commitment to democracy. Nevertheless, its location near the disembarkation point for immigrants made it easy to make the association. Some, including Thomas Bailey Aldrich, imagined the statue as a guard against the menace of undesirable immigrants. However, the more common connection with immigration was made in Emma Lazarus's poem, "The New Colossus," written in 1883 before the statue was erected. By 1903 the link between the Statue of Liberty and immigration was officially recognized, as words from her poem were inscribed on a plaque, bidding the world to send America its huddled masses, tired, poor, and yearning to be free. The plaque at first was placed not by the front door where it is today, but on the second floor, where it attracted less attention. There was no official installation ceremony. Acquiring meanings that Bartholdi had never intended, the statue was beginning to displace Columbia as the female personification of the nation.[7]

The Statue of Liberty held aloft an illuminated lamp whose radiance was exaggerated on postcards, but as its fame grew it was becoming a shadowy contrast to the brilliant New York skyline.[8] In response the *New York World* raised funds for 225 searchlights that bathed the statue in 20,000,000 candlepower.[9] On December 2, 1916, President Wilson and the French ambassador inaugurated this new system in a gala celebration where the skyscrapers turned on all their lights.[10] Both day and night, the statue became a vivid reference point for tugboats, ocean liners, ferries, and troop ships as they sailed for Europe. In the iconography of fund drives to sell war bonds, the Stature of Liberty held aloft the electrified torch of democracy.[11] One poster directed at recent immigrants depicted the statue as seen from the deck of a ship and carried the text, "Remember Your First Thrill of American Liberty: Your Duty—Buy United States Government Bonds." The press linked the statue to the struggle for victory over Germany and to Wilson's intention to make the world "safe for democracy."[12] By the 1920s the statue had become such a part of the American imagination that it appeared on a postage stamp for the first time, and it was routinely associated with heroic events, such as Lindbergh's flight to Paris (figure 6.1).

Yet the public did not permanently embrace every new monument. Some became a national symbol for a short time and then gradually faded into obscurity, as happened with President Grant's tomb in New York City. General Ulysses Grant led the Union Army to victory in the Civil War and was elected to two terms as president. Revered in life, when Grant died 90,000 people donated funds to build the largest mausoleum in the

Figure 6.1
"Remember Your First Thrill of American Liberty," Liberty Bond poster, Department of the Treasury, 1917. Courtesy of the Library of Congress, Prints and Photographs Division.

United States, where he and his wife Julia were buried. In 1897, one million people came to its dedication and parade. At first, his mausoleum attracted more visitors than the Statue of Liberty. Nostalgic Civil War veterans often came, but as they passed away the reverence for Grant attenuated. Located in uptown Manhattan at 122nd Street and Riverside Drive, the enormous granite and marble structure slowly fell into disrepair before restoration in 1997. It attracted far fewer visitors than the Lincoln Memorial in Washington.

While General Grant was celebrated in New York, the former Confederate states erected statues to Robert E. Lee and other generals who had fought to leave the United States. A central problem for the nation was to create symbols that southerners would endorse. Northerners therefore tolerated the mythology of the "lost cause" that flourished in Dixie. Nostalgia for the plantation South became widespread, abetted by the writings of the Dunning School of historians at Columbia University. As Eric Foner notes, they depicted slavery as a benign paternal institution and saw Reconstruction as "an orgy of corruption and misgovernment."[13] The Dunning School defined Black people not merely as uneducated but as mentally inferior. One of the most influential historians promoting such views was Ulrich B. Phillips. He received a PhD from Columbia in 1902 and taught at the universities of Wisconsin, Tulane, Michigan, and Yale. He argued that enslaved people were provided satisfactory housing, clothing, and food, and he had almost nothing to say about slave rebellions, runaways, or the psychological harm enslavement inflicted. Phillips believed enslaved people were less intelligent than their masters and saw plantations as Negro educational institutions. In *American Negro Slavery* (1918) and *Life and Labor in the Old South* (1929) slave agriculture was described as an often-unprofitable system that was maintained to enforce social control and keep the South "a white man's country." Richard Hofstadter later attacked "Ulrich B. Phillips and the Planation Legend," criticizing his heavy reliance on the records of a tiny minority of wealthy planters, with little attention to the far more numerous farmers who had only a few enslaved people. In his romanticized vision of large, well-run plantations with two hundred or more enslaved people, Phillips "chose to portray the Negro slave as a singularly contented and docile 'serio-comic' creature. His casual treatment of the enslaved person's resistance to slavery and of slave revolts was accordingly inaccurate." Likewise, Phillips ignored the subject of miscegenation and the testimony of southern physicians on the poor health of enslaved people.[14] His nostalgic history long remained unchallenged, and in 1944 Hofstadter was one of the first to expose his biased methods and selective use of sources.

During the first decades of the twentieth century, Phillips's views were accepted as orthodoxy. President Wilson imposed segregation at the White House, and he gave cabinet members free reign to segregate work in their departments. The federal government had been integrated since Abraham Lincoln was president, but in 1913 the post office and the treasury introduced segregated work areas and bathrooms. Other parts of the federal government followed suit, effectively denying Black employees a path to advancement that had existed until Wilson, a Virginian, was inaugurated.[15] Discrimination persisted in the Harding administration. When the new Lincoln Memorial was dedicated in 1922, the crowd was divided, with separate seats for whites and for Blacks.[16]

Not only was the plantation rehabilitated and depicted as a form of beneficent paternalism for happy but mentally deficient enslaved people, not only was segregation imposed first in the South and then on the federal government, but also the Civil War itself was reconceived. Michael Kammen summarized: "According to orations, sermons, editorials, and essays, each and every partisan had been brave, valiant, and courageous. Both sides had fought for causes they believed to be just. The history of racial issues in general, and the institution of slavery, virtually disappeared from mainstream public discourse."[17] The battlefields became memorial landscapes, notably at Gettysburg. A private association had begun to protect the site a few months after the battle. By the 1870s veterans from the North had begun to erect monuments, which became numerous in the 1880s. They focused on tactics, heroism, and sacrifice rather than on what the war had been about. In 1895 the 522-acre site, including more than 300 monuments,[18] was handed over to the federal government when Congress established the Gettysburg National Military Park. Much effort focused on marking the location of the battle lines held by Confederates and Union troops. Gettysburg was defined as the climax of the Civil War, and the climax of that battle was Pickett's Charge, a valiant but doomed attack upon the Union lines. Veterans came from both North and South, and the former enemies began to fraternize, especially on Independence Day, which coincided with the battle's anniversary. In 1888, a quarter-century after the battle, a large contingent of veterans came, and in 1913 more than 55,000 people celebrated the fiftieth anniversary. Edward Linenthal characterized the speeches on that occasion as being suffused with "the ideology of reconciliation." Both the "speeches and ritual activity emphasized collective heroism and the importance of healing from sectional strife." Pickett's charge was reenacted, and the conclusion was: "It was not Southern valor, nor Northern valor. It was, thank God, American valor."[19] President Wilson echoed these nostalgic themes in his address to the assembled veterans. He praised the vigorous growth of the reunited nation during the fifty years since the battle. "How wholesome and healing the peace has been! We have found one another again as brothers and comrades in arms, enemies no longer, generous friends rather, our battles long past, the quarrel forgotten—except that we shall not forget the splendid valor, the manly devotion of the men then arrayed against one another, now grasping hands and smiling into each other's eyes."[20] Wilson used the word "nation" twelve times, but he never referred to Lincoln, the Confederacy, Black Americans, equality, emancipation, slavery, abolition, or Reconstruction. His speech erased Black Americans. Wilson's Civil War was not about slavery and the struggle for freedom but about the valor of white men.

Erecting monuments and redefining the Civil War served to unify white Americans in nostalgic admiration for a heroic past. But there remained the disintegrative forces of industrialization that redefined work, destroyed many businesses, and upset class structures. As Jackson Lears notes, "myriads of thoughtful Americans," including graduates of the finest universities, "by the 1880s, had begun to question the very basis of industrial, capitalist society: not merely the unjust distribution of wealth and power but the modern ethic of instrumental rationality that desanctified the outer world of nature and the inner world of the self." Lears studied these dissenters who "recoiled from this ethic and groped for alternatives in medieval, Oriental, and other 'primitive' cultures."[21] This vocal minority of anti-modernists resisted secularization and the instrumentalist focus on efficiency. They sought spiritual rebirth and purification. The construction of memorials sought an anchoring point in the values of the founding fathers and the American Revolution. But the anti-modernists did not celebrate revolutionary values, for these values had helped to unleash the industrial capitalist economy they disdained.

Instead, a more distant past attracted the historian Henry Adams, who became fascinated by the medieval period and the cultural force of

Christianity. He became deeply interested in gothic churches, visiting many of them in France. Adams saw the cult of the Virgin Mary as a powerful antithesis to the modern faith in science and technology, and he expressed this polarity by contrasting the gothic cathedral with the electric dynamo.[22] Like Adams, many were so distressed by American industrial society that they became nostalgic for the pre-industrial European past. This strain of anti-modernism included an attraction to Catholicism, the revival of gothic architecture, and an appreciation for pre-modern paintings that were being gathered into enormous new museums in Boston, New York, Philadelphia, and other industrial cities.

Anti-modernists like the architect Ralph Adams Cram considered medieval art and design to be rooted in a living tradition that was superior to the secularism of their own times. Cram wanted to restore idealism through aesthetics, notably in gothic architecture for university buildings and in the majestic churches he designed for Boston, Cleveland, Detroit, St. Paul, Denver, Houston, and many other cities, notably the Cathedral of St. John the Divine in New York City. Cram's ideas ran counter to modernity, as he explained in *Walled Towns*. Published just after World War I, it declared, "We have seen how fragile, artificial and insecure is civilization, how instantly and hopelessly it can crumble into a sort of putrid dissolution." He feared a future in which there might be "a general breakdown of ideals, and either an accelerating plunge into something a few degrees worse than barbarism, with the Dark Ages as its inevitable issue, or an equally fatal return to the altogether hopeless, indeed the pestilential, standards and methods of the fruition of modernism in the world-before-the-war."[23] Cram wanted a revised version of the European medieval order. He rejected high-energy civilization in favor of water wheels and muscle power. Americans were rushing to buy automobiles, while Cram advocated building walled towns where they would not be permitted. Women had just won the right to vote, but he advocated the return to a patriarchal social structure. People were buying on credit as never before, but Cram wanted to abolish usury, because "the modern world has become a perfectly artificial fabric of complicated indebtedness, the magnitude and ramifications of which are so enormous that nothing preserves it but public confidence. Were this removed, or even shaken seriously, the whole fabric would collapse in universal bankruptcy."[24] The assembly line was rapidly spreading to thousands of factories, but Cram wanted to revive the guild system to govern artisanal production for durability and long-term use rather than mass production for profit. There were a great many religions in the United States, but Cram imagined the new walled towns would have a single unifying system of beliefs. In short, his anti-modernism sought to abolish the new energy forms and to reverse the trends toward secularism, gender equality, industrial production, mass consumption, individualism, and aesthetic eclecticism. Cram called for a wholesale rejection of the present in favor of a pre-industrial golden age, before the Reformation and before the rise of centralized national authority. He realized that an actual city wall was not necessary for an anti-modern community, but he thought it would provide a desirable architectural framework for its values.[25]

Cram had traveled extensively in Europe, as had many wealthy Americans. There were small colonies of expatriates in London, Paris, Florence, Vienna, and elsewhere. Many American artists and writers also lived abroad for long intervals, including Mark Twain, William Dean Howells, Henry James, Edith Wharton, James Whistler, Ernest Hemingway, F. Scott Fitzgerald, Gertrude Stein, T. S. Eliot, and numerous others. Many judged American culture to be immature, and the contrasts between the Old World and the United States were a common theme in novels and travel books. In 1879 James summarized this view:

> One might enumerate the items of high civilization, as it exists in other countries, which are absent from the texture of American life, until it should become a wonder to know what was left. No State, in the European sense of the word, and indeed barely a specific national name. No sovereign, no court, no personal loyalty, no aristocracy, no church, no clergy, no army, no diplomatic service, no country gentlemen, no palaces, no castles, nor manors, nor old country-houses, nor parsonages, nor thatched cottages, nor ivied ruins; no cathedrals, nor abbeys, nor little Norman churches; no great Universities nor public schools—no Oxford, nor Eton, nor Harrow; no literature, no novels, no museums, no pictures, no political society, no sporting class—no Epsom nor Ascot! Some such list as that might be drawn up of the absent things in American life—especially in the American life of forty years ago, the effect of which, upon an English or a French imagination, would probably as a general thing be appalling. The natural remark, in the almost lurid light of such an indictment, would be that if these things are left out, everything is left out.[26]

This sense of American cultural inadequacy, coupled with a preference for European culture, had attenuated somewhat by the 1920s. Yet in 1922,

Eliot's *The Waste Land* appeared, written in Europe and saturated with references to the European past. Like Adams and Cram, Eliot was drawn to the Catholic Church and the pre-modern world, and he disparaged the "unreal city" of modern London.[27] Lears took his title, *No Place of Grace*, from a line in Eliot's "Ash Wednesday."[28]

This critique of modern industrial culture was the prerogative of an educated minority. The middle class was more likely to seek a third alternative lifeworld where they could escape from the vertiginous onslaught of modernity. They were nostalgic for the American small town, understood as a repository of steady virtues. When asked where they would like to live, many Americans in the 1920s said they would prefer a small town. At the very moment when the central cities drew millions of people like a magnet, an idealized small town remained the iconic landscape of safety and comfort. Sheila Webb has argued, "The small town can be seen to nurture a feeling of community, to promote traditional values, to foster the practical attitudes that helped to create an entrepreneurial culture, and so serve as a launching pad for American ingenuity and progress."[29] Even literary works that satirized small towns, such as Sinclair Lewis's *Main Street* or Sherwood Anderson's *Winesburg, Ohio*, testified to the enduring hold such communities had on the popular imagination. Thorstein Veblen declared in 1923, "The country town is one of the great American institutions."[30] Randolph Bourne wrote in 1913, "An American town, large enough to contain a fairly complete representation of the different classes and types of people and social organizations, is a real epitome of American life." Bourne feared that cities would overwhelm such towns. "The vortex of the city, even the smaller city, is so powerful that it sucks in the hardiest and sometimes the most distant towns and strips them of all their individuality and personal charm. The city swamps its neighbors, turns them into mere aggregations of expressionless streets lined with box-like houses or shanties of stores, and degrades their pleasant meadows into parks and sites."[31]

The small town represented past virtues. As depicted in 1920s advertising, it was an idealized place "of unity, neighborliness, and comfortable human scale." Roland Marchand noted, however, that "few actual American villages of the 1920s and 1930s" resembled these images, but by comparison were rather "disjointed, slovenly, and discordant." The idealized town of advertising almost always contained a church steeple and a cluster of houses. The main street had a bank, stores, and a movie theater, but "grain elevators, mills and other evidences of processing or production for export were virtually non-existent." These images of "pristine, eternal villages" reassured the public that "despite the advance of awesome and impersonal skyscraper cities, their society still retained the qualities suggested by 'village America.'"[32] In many homes, an idealized miniature town much like that depicted in advertising was erected along the tracks of a model railroad in the basement or attic. And as Miles Orvell notes, during the 1920s the idealized small town was being recreated at full scale in Colonial Williamsburg, Virginia, paid for by Rockefeller money, and at Greenfield Village in Michigan, a Henry Ford project.[33]

Nor was this merely a nostalgic vision for the Fords and Rockefellers. Many ordinary Americans bought houses with a colonial revival style. Sears, Roebuck sold thousands of such prefabricated houses, some based on Washington's Mt. Vernon home, others in a New England style. Likewise, it seemed appropriate to many Americans that the president between 1923 and early 1929, Calvin Coolidge, was born on Independence Day, 1872, in the village of Plymouth Notch, Vermont. In 1923, Vice President Coolidge was there on vacation when he learned that President Harding had died. At 2:00 a.m., Coolidge took the oath of office in the family living room by the light of a kerosene lamp. When he died ten years later, Coolidge was buried in Plymouth Notch, and the village, consisting of just twelve buildings, was preserved as a memorial.

The poet Robert Frost likewise seemed to many Americans a representative of rural and small-town life. Born in California in 1874, after his father's early death the family returned to their New England origins. Frost graduated from Lawrence High School in Massachusetts and went to Dartmouth for one semester and to Harvard for three. He developed a distinctive poetic voice that drew on the colloquial speech of New England. Between 1916 and 1928 he published five books, won the Pulitzer Prize in 1924, and became one of the nation's most popular poets. In the public imagination, Frost was a rural sage whose poems offered what he once called "momentary stays against confusion."[34] He had an appealing persona and the world in his poems seemed a contrast to kaleidoscopic America. He did not overtly deal with urban life, industrialization, immigration, political ideology, or new technologies, but rather with rural settings and subjects. Writing on universal themes in accessible language, he produced poems that many felt they could understand. His work ignored the haste of the present in favor of timeless realities: the seasons, harvests, country walks, ice storms, fires, hard labor in the fields, and home burials. Frost's pastoralism provided a counterpoint to urban America. His work seemed an escape into rural life. Unlike the nineteenth-century romantic poets, however, Frost did not view the natural world as beneficent, and he could depict it as unforgiving or harsh. Yet Americans, disturbed by the disruptions of space, time, and identity, could read Frost as a modern Wordsworth. The popularity of small towns and of Frost's poetry exemplifies the idealization of the rural past in contrast to the frantic urban scene.

The first three examples of nostalgia focused on recuperating an earlier lifeworld. The construction of monuments to American values and founding fathers looked back a century or more; the anti-modernists longed for much earlier eras; and the idealization of small towns was based on the belief that traditional values were preserved away from the city. The fourth response was more radical, moving not only back in time but also to regions that were defined as being outside society. Even before the Civil War, the United States conceived of itself as "nature's nation." Lacking the hallmarks of European nationalism, such as Roman ruins, royalty, ancient cathedrals, or a long literary and artistic heritage, Americans instead saw the nation as an escape from history into untouched wilderness, where their character had been forged.[35] Alfred Runte explains: "The agelessness of monumental scenery instead of the past accomplishments of Western Civilization was to become the visible symbol of continuity and stability in the new nation."[36] If the United States did not have a Stonehenge, it had the Grand Canyon. If it lacked the ruins of empires, it had ancient redwood trees in California that were already growing when Augustus Caesar ruled Rome.

In short, Americans became nostalgic for what they conceived as untouched nature. Certain sites became the quintessential natural landscapes, and they were protected as national parks. The first was established at Yellowstone in 1872, and most early parks were in the western half of the United States. These sites had no associations with the Civil War, and it was easier for a former Confederate to identify with Glacier National Park in Montana or the Grand Canyon in Arizona than it was to identify with any site in the Midwest or North. In 1879 the federal government established the Geological Survey as a separate bureau within the Department of the Interior. It surveyed lands in the national domain, particularly in the western states and territories. The surveys identified places that might be exploited and areas that might be set aside. The national parks movement recognized and protected pristine areas that could be overrun by logging, mining, and industry. Parks were to preserve timeless remnants of wild nature. Both white settlers and Native Americans were evicted from these fragments of "original America," which afterward had no permanent residents except for government caretakers. Yosemite or Yellowstone once had been inhabited, but visitors got the impression that explorers discovered them in an empty wilderness.

The western railroads promoted the parks, built their lines close to them, erected rustic hotels near their entrances, and employed artists to depict the splendid scenery.[37] The parks became an integral part of the national kaleidoscope, and each one had a trademark site, such as Old Faithful at Yellowstone, El Capitan in Yosemite, or the Grand Canyon viewed from Point Sublime. The journey to natural sites also was celebrated. One railroad in Pennsylvania advertised that "it is built along ravines so narrow that its bed is carved out of the overhanging rock. Now a mountain-spur bars the path and is pierced by a tunnel, and the river is so tortuous that numerous bridges carry the track from bank to bank. Every mile opens new scenes, which present themselves to the traveler's eye like the changing pictures of the kaleidoscope."[38] Likewise, the National Park Service described the chasm of the Yellowstone River framed by blue skies and surrounding green pines as being "colored as brokenly and vividly as the field of a kaleidoscope. The whole is streaked and spotted and stratified in every shade from the deepest orange to the faintest lemon, from deep crimson through all the brick shades to the softest pink, from black through all the grays and pearls to glistening white."[39] Yosemite was also appreciated as the epitome of the natural sublime. In 1915, a visitor declared it was "at the limit of one's capacity to grasp and comprehend. The whole valley is kaleidoscopic; a pen cannot depict nor can brush portray the subtle charm of its ever-shifting scenes."[40] Drawn by such rhetoric, enticed by stereographs, postcards, advertisements, and paintings, and driven by national pride, Americans flocked to the national parks, especially after families bought automobiles. Yellowstone Park had almost one million visitors a year by the end of the 1920s. Roughly half came by train and half in cars. Those who did not visit the parks saw them in postcards, magazine articles, and newspapers.

As interest increased, better administration became necessary. In August 1916, President Wilson signed the act that created the National Park Service to manage the thirty-five established national parks and monuments. Stephen Tyng Mather, who had made a fortune by mining and refining borax, was appointed the first Superintendent of Parks and served until 1929. He unified the administration. Individual parks had been run by military officers or political appointees without clear policies or goals, and some parks suffered from mining, logging, hunting, and commercial exploitation that was supposed to be outlawed. Some were overrun by grazing cattle, and private individuals ran small businesses inside them that catered to tourists. Mather transformed this haphazard administration into a systematic organization. By the time he retired, he had also convinced Congress to create seven more parks: Grand Canyon, Acadia, Bryce, Zion, Lassen, Hawaii, and Mount McKinley.

The back-to-nature movement included more than the national parks. State and private organizations also set aside nature preserves. In the East, the Appalachian Mountain Club, founded in 1876, became one of the most influential conservation groups. It preserved eastern wilderness areas, notably a large, sparsely settled region in northern New York State that bordered on Canada. In the West, John Muir founded the Sierra Club in 1892, which focused on protecting wild nature, particularly in California. At the same time, the new institutions of the Boy Scouts and the Girl Scouts gained popularity. Millions of children were sent to "wilderness" summer camps,[41] where they were encouraged to be self-reliant and live simply outdoors. They also acquired a romanticized view of Native Americans, but they had a "white" experience of nature, since other races were seldom included.

Theodore Roosevelt exemplified the growing appreciation of "roughing it" in primitive nature. After growing up in New York and attending Harvard, he lived in the Dakota Territory for several years. There he dressed like a cowboy, hunted wild game, lived in a cabin, and spent his days on horseback. Back in New York, Roosevelt wrote a six-volume history of the "winning" of the American West.[42] Likewise, in Frederick Jackson Turner's writings, the West became the region that best represented the nation, fostering individualism, courage, and self-reliance. Another example of this idealization of primitive nature was the erection of thousands of log cabins as vacation retreats.[43] The log cabin also was the inspiration for a rough-hewn architectural style used in buildings at both state and national parks[44] (figure 6.2).

Figure 6.2
Log cabin–inspired interior of the El Tovar Hotel, Grand Canyon, 1908. Stereograph, photographer unknown. Courtesy of the Library of Congress, Prints and Photographs Division.

To summarize: To cope with the rapid and confusing changes of the energy transition, Americans adopted four different forms of nostalgia to idealize the past without arousing the passions of the still divisive Civil War. One practice was to erect monuments to heroes of the Revolution, notably George Washington, or monuments to shared values, such as the Statue of Liberty. Furthermore, as illustrated by Wilson's speech at Gettysburg, Civil War battlefields and monuments were preserved not to recall the abolition of slavery but to celebrate the bravery and heroism of both Union and Confederate troops. A second approach was adopted by a minority of educated Americans, notably by writers, artists, and architects, who idealized not the United States but pre-industrial Europe. Third, many Americans became nostalgic for the small town, defining it as the epitome of a democratic community, in contrast to the city. Celebrating such towns could be harmonized with the fourth form of nostalgia, which identified with nature, whose sublimity was preserved in state and national parks. As the director of the Geological Survey mused, "Where better than in these wide spaces telling of the infinite past can tired and troubled humanity find relief from the petty things of the city street?"[45]

Each of these imaginary pasts was inscribed in the national landscape at tourist sites. Americans could fondly recall the early national period by visiting the Washington Monument, Bunker Hill Monument, the restored homes of George Washington and Thomas Jefferson, and other leaders of early America. They flocked to the Statue of Liberty, Grant's Tomb, Lincoln Memorial, and Civil War battlefields. They toured new gothic cathedrals and the classical architecture of college campuses erected by anti-modernists. Beginning around 1870 "Victorian gothic" buildings went up at Yale, Harvard, Cornell, the University of Pennsylvania, and Trinity College. Many more followed after 1880. The University of Chicago erected a copy of the Magdalen Tower of Oxford University. Ralph Adams Cram designed a gothic-style graduate school for Princeton (figure 6.3). The University of Pittsburgh merged the gothic revival with the skyscraper in a forty-two-story "Cathedral of Learning." Its hidden frame was of steel, though its cladding was limestone (figure 6.4). Some small towns also became iconic destinations. These included not only Rockefeller's restoration of Williamsburg but also Woodstock, Vermont; Concord and Stockbridge in Massachusetts; Saratoga Springs in New York; Annapolis, Maryland; Marietta, Ohio; Decorah, Iowa; Savannah, Georgia; and many other communities. Finally, national parks attracted millions. While idealized as unspoiled nature, they were accessible by railroads and automobiles, and their hotels boasted modern amenities such as hot and cold running water, central heating, and electric lights.

Figure 6.3
Gothic revival buildings of the graduate school of Princeton University, Ralph Adams Cram, architect, 1913. *Architectural Record* 35, no. 1 (January 1914): 3.

Figure 6.4
The Cathedral of Learning, Pittsburgh, circa 1925. Theodor Horydczak, photographer. Courtesy of the Library of Congress, Prints and Photographs Division, Theodor Horydczak Collection.

These nostalgic escapes from modernity demanded high energy consumption. Regardless of which idealized past tourists sought, they needed a car or a train ticket and a hotel. By the 1920s automotive tourism brought with it a cluttered landscape of billboards, flashing lights, gas stations, car dealers, drive-in restaurants, coffee shops, diners, bars, hot dog stands, souvenir shops, miniature golf courses, and other amusements.[46] The kaleidoscopic attractions—whether a monument, gothic revival building, exemplary small town, or national park—were embedded in commercial culture.

7

INDIVIDUALISM

Individualism has long been an important part of American identity, closely associated with economic independence and self-reliance.[1] The word "individualism" came into the American language when Tocqueville's *Democracy in America* was translated into English. He had coined the term to describe what he found distinctive in the United States, and he defined individualism in contrast to selfishness. "Our fathers were only acquainted with egoism (selfishness). Selfishness is a passionate and exaggerated love of self, which leads a man to connect everything with himself and to prefer himself to everything in the world. Individualism is a mature and calm feeling, which disposes each member of the community to sever himself from the mass of his fellows and to draw apart with his family and his friends, so that after he has thus formed a little circle of his own." Tocqueville related this less selfish strain of egoism to the abolition of royalty, aristocratic titles, and inherited class positions. "As social conditions become more equal, the number of persons increases who, although they are neither rich nor powerful enough to exercise any great influence over their fellows, have nevertheless acquired or retained sufficient education and fortune to satisfy their own wants. They owe nothing to any man, they expect nothing from any man; they acquire the habit of always considering themselves as standing alone, and they are apt to imagine that their whole destiny is in their own hands."[2] The equality of all citizens encouraged individualism.

The young nation that Tocqueville visited during 1831 and 1832 was primarily agrarian, without many large industries. Railroads had not yet been built; the national government was small; and the white population was rapidly settling the West from Ohio to the Mississippi Valley. In contrast, after 1876 Americans were concentrating in cities where class distinctions became more pronounced. Urban culture was rapidly displacing rural life, and small workplaces were being subsumed by corporations. The overarching narrative was that society progressed as it mastered new technologies,[3] and at an abstract level this was appealing. As Michael Adas has demonstrated, the ability to create, develop, and use new machines was thought to be a measure of Western society's superiority over what were perceived as the slothful, imprecise, and inefficient peoples of Africa and Asia. But how did one write the narrative of industrialized progress on a personal level? The price of rising energy intensity seemed to be submission to the machine. Could individualism be preserved? Homesteading had still been possible in some parts of the West in 1876, but it was rare by 1900. Indeed, Frederick Jackson Turner had declared the end of frontier America at the Chicago Exposition, in a celebrated address that set the agenda for a generation of historians.

By 1920, Turner realized that the preservation of individualism in an industrial society was not assured. He still argued that "frontier individualism has from the beginning promoted democracy," but he concluded that unrestrained individualism had become a problem as the economy matured: "the farmers of the Mississippi Valley gradually learned that unrestrained competition and combination meant the triumph of the strongest. . . . They learned that between the ideal of individualism, unrestrained by society, and the idea of democracy, was an innate conflict; that their very ambitions and forcefulness had endangered their democracy." Turner concluded that this conflict led to the revolts of the Grangers, the Greenback movement, the Populist Party, and Roosevelt Republicanism. Turner expected that American individualism would be modified to accept regulation, that it was being strengthened through state-funded universities, and that the nation was "engaged in the task of readjusting its old ideals to new conditions and is turning increasingly to government to preserve its traditional democracy."[4]

Yet many of Turner's contemporaries did not think individualism was worth preserving. In 1906, when President Roosevelt was breaking up some of the largest monopolies, *The Outlook* magazine declared, "We have passed out of the era of industrial individualism and unrestricted competition into that of industrial combination which inevitably tends to industrial monopoly," which "is wholly inconsistent with human liberty and is not to be endured by a free people."[5] Such views were common among social reformers and were taken seriously by sociologists and political scientists. The widely used textbook by Henry Reed Burch and S. Howard Patterson, *Problems of American Democracy*, argued that "after a century of experiment, laissez faire or individualism has been tried in the balance and found wanting. The world is now in a period of transition toward another stage of social development which we may call that of social welfare."[6]

Many Americans did not accept such predictions, however. Like Turner, they wanted to preserve the ideals of self-reliance and individualism. Furthermore, as Rebecca Wright argues, some progressives believed the challenge of their age was to direct and control energy through willpower. William James's "philosophy of affirmation foregrounded the will as a harnesser of energy," and his later works praised "the heroic individual able to act on the world and bring order into existence." His student, Theodore Roosevelt, "spent the best part of two decades (as both president and national celebrity) promoting the will as a mechanism to overcome the gradual siphoning off of energy." He promoted conservation, efficiency, the strenuous life, and self-creation, at times in support of war, to the disapproval of his former teacher.[7]

The average American was less likely to read William James than the Western novels that celebrated the individualistic figures of the scout, the trapper, the cowboy, and the frontier farmer. Owen Wister's *The Virginian* showed the appeal of the Western genre, and it was the best-selling novel of 1902. Harold Bell Wright's *The Winning of Barbara Worth* (1911) depicted the travails of settlers who struggled to divert Colorado River water to irrigate the Californian desert. The story of how they transformed arid land into "The Imperial Valley" was a bestseller and later a feature film.[8] The popularity of Western novels was further confirmed by Zane Grey's works, such as *Riders of the Purple Sage* (1912). Even if the homesteading frontier was gone, many still identified with a strong-willed hero and imagined the West as the land of economic opportunity.

Yet in practice large-scale farmers rather than homesteaders reaped many of the advantages of irrigation, and mining soon shifted from individual claims worked with picks and shovels to corporations using expensive machinery.[9] It took capital to build irrigation dams, install electric generators, erect transmission lines, and operate pumping stations. The larger mines installed electric lighting and cut haulage costs in half when squat electric locomotives replaced men driving mule carts. As in the city, success and profits accrued to those who commanded advanced machinery. There was a general recognition that the individualistic frontier had closed, marked not least by Buffalo Bill's death in 1917. There seemed to be little room for individualism in the vast irrigated farms, corporate mines, or railroad towns, any more than in large corporations, assembly line factories, offices filled with white-collar workers, or the crowded streets of American cities.

Nevertheless, many were convinced that opportunities abounded in industrial society. The popular orator Russell Conwell spread this message to all parts of the nation in a speech he delivered more than 5,000 times and turned into a best-selling book. He proclaimed that "acres of diamonds" were waiting to be discovered right under the noses of ordinary Americans. Conwell declared that "you ought to get rich, and it is your duty to get rich." He told audiences about a poor man who started with nothing more than his penknife and some firewood. He whittled wooden toys for his children, realized other children might like them too, started a business to do just that, found a good market for his wares, and became wealthy. Or there was an ordinary woman who was now rich because she invented a new kind of button that snapped closed, or the man who became rich after he manufactured the first pencil with an eraser attached. Wealth came to those who carefully examined the world around them and discovered opportunities. The examples varied, but Conwell's repeated message was that profitable inventions were made by those who discerned what people wanted, whether an improved button, a wooden toy, a sewing machine, or a pencil with an eraser. Success was not the result of genius or inherited wealth. "The really great man is a plain, straightforward, every-day, common-sense man."[10] Conwell was not only an orator but also a minister of a large Philadelphia congregation, the founder of a hospital, and the founder of Temple University, dedicated to the education of working-class students. To them all, he incessantly preached the gospel of success through self-help.

Athletics offered another avenue to fame and fortune. By the early twentieth century, Americans were celebrating professional sports as a new arena for individualism. Football was immensely popular, and baseball became the national game. Roderick Nash noted that in the 1920s athletes seemed to exemplify "the power of courage, strength, and honor" and "the efficacy of the self-reliant individual." The "sports fan regarded the playing field as a surrogate frontier" and "the athletic hero was the twentieth century equivalent of a pathfinder or pioneer."[11] The sports star often came from a humble background, worked hard to make the most of an inherent talent, and was already famous by the age of twenty-five.

Athletic heroics were part of a growing emphasis on the strenuous life and masculinity, epitomized by Theodore Roosevelt. Born to East Coast wealth, he transformed himself into a cowboy in the Dakota Territory. When he later raised a troop called the Rough Riders to fight in the Spanish-American War of 1898, Roosevelt rode into the national limelight. Before then his books on the American West united frontier individualism with military valor. In both western conquest and the war that wrested Cuba and the Philippines from Spain, there was a strong element of racial conquest. At its worst, such attitudes led to brutal extermination of Native Americans at Wounded Knee in 1890, or to guerilla war a decade later in the Philippines. Like Reverend Strong, many believed that the Anglo-Saxons were destined to conquer other races.[12] There also was a military undertone to physical education that became mandatory at the universities at the same time that intercollegiate sports emerged. As Lears notes, by the 1880s "a new emphasis on physical vitality" became common and "waves of muscular Christianity began to wash over campuses."[13] Universities built stadiums and celebrated student athletes. By the 1920s, college football games were major events.

In short, after the passing of the frontier, the mythology of individualism remained strong in various guises. Moreover, the gradual rise in wages, together with shorter hours, suggested that there was general improvement in the quality of life. As discussed in chapter 2, after 1880 there also was a shift from blue-collar to white-collar occupations, as well as a rapid increase in the number of patents. Manufacturing new products, new corporations emerged to dominate the economy. Indisputably, during these developments some people had risen from modest backgrounds to great wealth and fame, notably, Andrew Carnegie and Henry Ford. Their success seemed to prove that the individual, freed from the constraints of class, tradition, and political control, could benefit society by pursuing self-interest.

Benjamin Franklin's *Autobiography* was still widely read, and it provided a model for such individualism. In the eighteenth century, Franklin broke away from his family and his apprenticeship in Boston and went to Philadelphia with nothing more than he could carry. There, he became a successful printer and made important scientific discoveries. He became a civic leader, the representative of several colonies in London, a leader of the American Revolution, ambassador to France, and, finally, one of the architects of the Constitution. These remarkable achievements were based on self-education and hard work, combined with personal charm. Franklin's story became for later generations the essential promise of American life:

Anyone could succeed because the United States had eliminated royals and aristocrats to create a level playing field and offer equal opportunities to all. His autobiography was frequently reprinted, in full, in part, and in abridged children's editions. It was obligatory reading at many schools. Between 1880 and 1900 alone there were eleven new printings from American publishers.

Andrew Carnegie saw himself as a latter-day Franklin. He distributed thousands of copies of Franklin's *Autobiography*, as well as building more than 2,500 libraries in the English-speaking world. Most of these were in the United States, to ensure everyone had access to knowledge. The son of a poor Scottish immigrant to Pittsburgh, Carnegie worked his way up in the steel industry from a delivery boy to the richest man in the world. He was ruthless in adopting new industrial methods in his drive for efficiency, and he fought the unions. Eventually, he controlled the nation's steel industry and then sold it in 1901. For the rest of his life Carnegie spent his fortune on philanthropy. As he explained in a widely republished essay, "The Gospel of Wealth,"[14] children of the rich should not inherit a fortune; it should be devoted to the community. Carnegie not only built libraries, but also founded a university in Pittsburgh, created an award for heroism, and paid for construction of the World Court in the Netherlands. Carnegie famously declared of those who kept their wealth rather than using it for philanthropy, "He who dies thus rich dies disgraced."[15] Such ideas influenced other capitalists to become philanthropists, usually focusing not on direct payments to the poor, but on institutions such as hospitals, schools, training programs, or charities.

Herbert Hoover was another example of a poor boy who made good. Orphaned as a child, he rose from modest beginnings in rural Iowa to study at Stanford University and then made a fortune as a mining engineer, before going into politics. In 1922, Hoover published *American Individualism*, which linked the pioneers of the nineteenth century with the opportunities created by an industrial economy: "Our American individualism has received much of its character from our contacts with the forces of nature on a new continent. It evolved government without official emissaries to show the way; it plowed and sowed two score of great states; it built roads, bridges, railways, cities; it carried forward every attribute of high civilization over a continent." Hoover had much the same message as Conwell: "The days of the pioneer are not over. There are continents of human welfare of which we have penetrated only the coastal plain. The great continent of science is as yet explored only on its borders, and it is only the pioneer who will penetrate the frontier in the quest for new worlds to conquer. The very genius of our institutions has been given to them by the pioneer spirit."[16] Hoover's ideas harmonized so well with those of Ford, Edison, and the tire manufacturer Harvey S. Firestone that he joined them for a weeklong motorized ramble in the countryside (figure 7.1).

To reinforce their individualism and pioneer spirit, many Americans embraced "positive thinking," which included mind cures, "the power of thought," Mary Baker Eddy's Christian Science, New Thought, and other popular theories. Such books reached a wide audience, explaining "the healing power of mind" and "what shall make us whole," as two book titles proclaimed. A best-seller by Ralph Waldo Trine, *In Tune with the Infinite*, promised that mind cure could deliver peace, power, and plenty. This movement explored the region between religion and psychology, claiming to have found techniques to align the subconscious mind with outside forces

Figure 7.1
Henry Ford, Thomas Edison, Herbert Hoover, and Harvey S. Firestone on vacation. Harris & Ewing, photographers, 1922. Courtesy of the Library of Congress, Prints and Photographs Division.

that could cure mental and physical ailments. It seemed that these techniques also could release energies that ensured personal success.[17] Henry Ford was attracted to such ideas, and he told a mystic visiting from India in 1926 that "if I quietly withdraw from the nervous anxiety over things, inventions and the business that drives from every side, there was a renewal of strength in the thought of being a part of the unseen power."[18] Ford once sent Douglas Fairbanks a copy of Trine's *In Tune with the Infinite*, thinking it might help him to complete a difficult film. In the late 1920s Trine and Ford published a dialogue between themselves with the title *The Power That Wins*. It promised readers "an intimate talk on life—the inner thing—the things of the mind and spirit—and the inner powers and forces that make for achievement." Ford averred, "Anybody can do anything that he imagines," and he discussed such ideas with his friend Thomas Edison.[19] However, Carnegie, Rockefeller, and many other heroes of individualism did not take much stock in mind cures.

Americans also knew of hard-working laborers who had failed, and their stories appeared to contradict both the dominant narrative of individualism and the hopeful claims of mind cure. In popular culture, failures were often explained in cautionary tales about the terrible consequences for those who could not resist urban temptations. *Sunshine and Shadow in New York* typified this genre's stress on the need for a strong moral character. "Bad men locate in New York, fix there their headquarters, and reduce roguery to a system. They have their banks, expressmen, artists, and agents. These men dwell in the dark recesses and hidden chambers of the city." They laid in wait. At the same time, "to New York come also the most talented and best of men. The talent, ability, integrity, shrewdness, and sharpness which made a small fortune in any other place make a large one in New York."[20] The book alternated between descriptions of stunning success and abject failure, with chapters on wealthy men such as William Astor or Cornelius Vanderbilt interspersed with chapters on "Black Mailing as an Art" or "Business Reverses in New York." The book described many roads to ruin, including fraud and embezzlement. "Dishonesty, stock gambling, expensive horses, with women, wine, fast and high living, tell the story." In defeat, "many have given up hope, and taken to the bottle. New York is full of wrecks of men who, because they could not pay their notes, have flung away character, talent, and all."[21] The city destroyed many: "young men come in from the country, confident of success, and ignorant that these [successful] men are the exceptions to the general law of trade; and that ruin and not success, defeat and not fortune, bankruptcy and not a fine competence, are the law of New York trade."[22] The successful had avoided urban temptations. Failure was rooted in moral weakness. Other popular works such as *The Secrets of the Great City* had a similar message, describing "the virtues and the vices, the mysteries, miseries, and crimes of New York City." It praised Central Park, the hotels, the theaters, and fashionable shopping, but also devoted sections to imposters, swindlers, hotel thieves, gamblers, beggars, blackmailers, sex workers, pickpockets, pawn brokers, and other characters. The book's final sentence warned readers to avoid the "darker quarters of the city," which it had described in alluring detail, because they were "gateways to the paths that lead down to ruin and death."[23]

The city was often described as a cesspool of corruption. An article in the *Fort Worth Star-Telegram* described the tragic death of a young man in the "red-light district" as "the result of burning the candle at both ends," which ought to "be a lesson to every young man in the city." The dead man had a good upbringing but was led astray by the "fast life." Such a young man "chuckles with pleasure at the dazzle of the life by which he is attracted. The child peering into the kaleidoscope sees the pretty colors only," and similarly the "young man watching the changes of the colors in the kaleidoscope of life sees only the brilliancy and looks not beyond. He follows the brilliancy" until "he is lost to shame, to dignity, to refinement and to morality."[24] The city's temptations could destroy a man through ill-considered investments, alcohol, gambling, prostitution, and extravagant living. Again, failure revealed a weak character.

If the city was rife with temptation, nature offered every man a fresh start, exemplified not only by the frontiersman but by the fictional hero Robinson Crusoe. His story was immensely popular in nineteenth-century America, not only in Daniel Defoe's original version but in many imitations. By 1900 more than 150 editions and abridgments of Defoe's novel had appeared in the United States. Its imaginary triumph of individualism was one of the most widely read stories of the age.[25] Isolated on an uninhabited island, Crusoe was in sole control of his fate. Without assistance, he built a fortified dwelling, domesticated animals, developed agriculture, and created a viable existence. His story showed readers that a castaway could master adversity. Who embraced this story? The millions of immigrants arriving with no more than they could carry, the young men and women leaving farms to try their luck in the burgeoning cities, the fundamentalist Christians who could see Crusoe's shipwreck as a return to the Garden of Eden where he recovered his faith, the capitalists such as Andrew Carnegie, and those who dreamed of going west and starting over.

When President William McKinley was a boy, two of his favorite novels were *Robinson Crusoe* and the somewhat similar *Swiss Family Robinson*.[26] Likewise, in 1881 a biography of President James A. Garfield revealed that when a schoolboy, he borrowed a copy of *Robinson Crusoe* and read it before the family fireplace. He was so enthralled that when he had finished the last page, he immediately read it again. He told his mother, "I never read such an interesting book. I could read it ten times over and not get tired of it."[27] He was inspired by Defoe's novel and became one of the better educated presidents. At Williams College, Garfield studied science and learned German and Hebrew. Throughout his career, he promoted education as being essential to democracy and progress.

The Crusoe story became the basis for a genre, the "robinsonade." For example, in 1837 Lucy Ford published *Female Robinson Crusoe: A Tale of the American Wilderness*, in which a young girl of eight flees a wolf and gets lost in the vast forests of the West. She manages to survive by building a rude shelter, gaining the use of fire from a lightning strike, and eating berries, nuts, and turtles. She accidentally discovers a trove of buried gold and silver coins, as well gardening tools, household utensils, iron pots, and an axe. Armed with these tools, she becomes self-reliant. "At home, I had been accustomed to a thousand clogs upon my freedom of thought and action, from which, here in the woods alone, and unobserved, I was entirely exempt." In the wilderness, "unchained and released from fixed times and tasks, I roved free as the air."[28] Her solitude ends when a lost boy she knows discovers her campfire, and together they establish a viable, self-sufficient household.

The Encyclopedia of Science Fiction defined the robinsonade as "the romance of solitary survival in such inimical (though ultimately compliant) terrains as desert islands (or planets), seen as a success-story." The central theme of these stories is the "celebration of the power of pragmatic Reason, and its depiction of the triumph, alone, over great odds, of the entrepreneur who commands that rational faculty."[29] There were many such stories.[30] As the nation became urban and interdependent, a self-sufficient hero was appealing. American protagonists often overcame more daunting obstacles than Defoe's hero. In 1885, the preface to Douglas Frazer's *Perseverance Island; or,*

The Robinson Crusoe of the Nineteenth Century, declared: "In all works of the Robinson Crusoe type, the wreck is always near at hand, the powder dry and preserved, and the days for rafting the same ashore calm and pleasant." But the fictional William Anderson "had no such accessories; and his story proves the limitless ingenuity and invention of man, and portrays the works and achievements of a castaway, who, thrown ashore almost literally naked upon a desert isle, is able by the use of his brains, the skill of his hands, and a practical knowledge of the common arts and sciences, to far surpass the achievements of all his predecessors, and to surround himself with implements of power and science utterly beyond the reach of his prototype, who had his wreck as a reservoir from which to draw."[31] Anderson builds a stone house, develops agriculture, makes gunpowder, mines iron ore, forges steel, creates a machine shop, builds a submarine, discovers a sunken pirate ship laden with gold and silver, and makes a balloon-assisted flying machine. He far outperforms Crusoe.

The story of *Robinson Crusoe* itself was partly based on the experiences of a British sailor, who spent four years and four months marooned off the coast of Chile before he was rescued.[32] The story has been seen as a parable about the origin of society, about the superiority of Western civilization, about John Locke's theory of property rights, and about the nature of economic man.[33] A free-market reading of Crusoe has appeared in many economics textbooks.[34] Michael V. White's classic essay, "Reading and Rewriting: The Production of an Economic Robinson Crusoe," explains how during the nineteenth century Defoe's novel was reconceived in terms of liberal microeconomics.[35] Crusoe began to be seen as the exemplar of economic man in the ultimate free-market economy with no government regulation.[36]

Defoe's contemporary, John Locke, had argued that in a state of nature a person acquires ownership of land by laboring on it. By that standard, Crusoe acquired the right to his island after he improved it. Once Crusoe achieved a comfortable life there, he realized that there was no point to raising more animals or growing more food than he could consume. He "found that the forty bushels of barley and rice was much more than I could consume in a year; so, I resolved to sow just the same quantity every year that I sowed the last." He had "enough to eat and to supply my wants, and what was all the rest to me? If I killed more flesh than I could eat, the dog must eat it, or the vermin. If I sowed more corn than I could eat, it must be spoiled."[37] Salvaged bars of gold from a shipwreck were of less value to him than an iron fire shovel and tongs.[38] He would have exchanged the gold for a hand mill to grind his corn or for a bottle of ink.[39] Crusoe had a farm, tame goats, a fortified home, and more food than he needed. He realized that "the most covetous griping miser in the world would have been cured of the vice of covetousness if he had been in my case, for I possessed infinitely more than I knew what to do with."[40] In the middle sections of the novel, he might seem a precursor of Henry David Thoreau or modern environmentalists.[41] However, most Americans focused on Crusoe in the first part of the book, when he rises from being a castaway to living a lonely, but comfortable life. His story appealed to Garfield, McKinley, and other Americans as the epitome of self-reliance, like a lone frontiersman opening a new region to settlement.

Defoe himself had been a merchant, and he wrote a pamphlet that praised the British patent system and "projectors" (or entrepreneurs).[42] His novel later seemed to exemplify the imperialist argument, reprised by Adas in *Machines as the Measure of Men*, in which technological mastery provides an index of each culture's level of civilization.[43] These imperialist implications were explored in Mark Twain's *A Connecticut Yankee in King Arthur's Court* (1889). His protagonist, a mechanic named Hank Morgan, wakes up after a concussion a thousand years back in time. He finds medieval England primitive, with poor lighting in drafty buildings. "There were no books, pens, paper or ink, and no glass in the openings they believed to be windows. It is a little thing—glass is—until it is absent, then it becomes a big thing. But perhaps the worst of all was, that there wasn't any sugar, coffee, tea, or tobacco. I saw that I was just another Robinson Crusoe cast away on an uninhabited island, with no society but some more or less tame animals, and if I wanted to make life bearable, I must do as he did—invent, contrive, create, reorganize things; set brain and hand to work, and keep them busy. Well, that was in my line."[44] Morgan tried to transform medieval Britain, and he became politically powerful as "the Boss." His innovations did not lead to democracy. He became a dictator, and the technologies he introduced led to a disastrous civil war. However, most of Twain's contemporaries believed that although social classes persisted, the life of the average person continued to improve. Advances in technology seemed to create greater efficiency and prosperity for all. They believed, with Conwell, Ford, Carnegie, and Hoover, that there were still vast opportunities in America.

Crusoe had the attributes of a skilled mechanic or inventor, and many Americans celebrated inventors as ideal citizens. The inventor was conceived as someone who worked alone, or with a small team, to create something that enriched themself and benefited society. The pantheon of such heroes included Eli Whitney (the cotton gin), Cyrus McCormick (the mechanical reaper), Samuel Morse (the telegraph), Alexander Graham Bell (the telephone), and the Wright Brothers (the airplane).

Female and Black inventors seldom received equal attention. As Ruth Oldenziel demonstrates in *Making Technology Masculine*, at the great world's fairs, at the Smithsonian Institution, and in the patent office itself, invention was considered to be a manly activity, and the most prestigious technologies were mechanical devices. If a women invented a device useful in sewing, housework, or elsewhere in the domestic sphere, its importance was automatically discounted.[45] Josephine Cochrane invented a dishwashing machine patented in 1886 that was used in many hotels and restaurants. At the Colombian Exposition she won a prize for its design and durability, but she was seldom celebrated or remembered.[46] Maria Beasley had fifteen inventions, including a life raft used on ships, and a barrel-making machine that netted her a large income. Yet she, too, was scarcely known.[47] Likewise, Black inventors were seldom lionized, in part because many had agricultural patents that were considered less prestigious than industrial patents. Henry Blair, a free Black man in Maryland, received patents for two seed planting machines in the 1830s, but his contributions were forgotten. George Washington Carver was the only Black inventor mentioned with any regularity. Born into slavery, he became a botanist who discovered hundreds of uses for peanuts, sweet potatoes, and other crops.[48]

The pantheon of invention was primarily male and white. As late as 1973, Richard Nixon declared in a message to Congress that "America's dramatic progress from a small agrarian nation to a great technological and industrial leader has been due, in no small degree, to the inventive genius of its people. Names such as Benjamin Franklin, Eli Whitney, Cyrus McCormick, Thomas Edison, Alexander Graham Bell, Samuel Morse, the Wright Brothers and Henry Ford speak volumes about the character of our Nation."[49] Of these, Edison was the most famous. Between circa 1870 and 1931 he was granted more than 1,000 patents, including the phonograph, the incandescent electric light, improvements to the telephone and the microphone, the practical motion picture camera and projector, and many more. Long before his death, a simplified story loosely based on his life had become a staple of juvenile fiction, and there was a tendency to cast the lives of other inventors in the mold of this idealized story. Eventually, the term "edisonade" was coined to describe this genre, derived from Edison's name just as "robinsonade" was derived from Robinson Crusoe.[50] The edisonade features a young inventor who rises from humble beginnings to fame. His ingenious contrivances extricate him from difficulties, defeat his opponents, make him wealthy, and enable explorations and adventures.

The edisonade simplified a welter of contradictory stories that were widespread during the first decades of Edison's fame. He had not even gone to high school, but many newspapers called him a scientist or professor, including the *New York World* and *New York Sun*.[51] Edison's ideas were also ridiculed. For example, the English scientist William Henry Preece declared it was theoretically impossible for Edison to subdivide the electrical current.[52] Others described Edison as a magician. *Harper's Weekly* called him "a midnight workman with supernal forces." He was "a modern alchemist, who finds the philosopher's stone to be made of carbon," a reference to his discovery that carbon could make a good filament for the electric light. Edison seemed to be a sorcerer, who "with his magic wand changes everyday knowledge into the pure gold of new applications." He was "conjuring Nature" and exploring mysterious worlds.[53] The *New York Herald* declared that "invisible agencies are at his beck and call," and reported that Edison had a skull on his desk and read strange books with "cabalistic characters."[54] Another version of Edison was the tinkerer or Yankee mechanic, a particularly skilled artisan who had no occult practices but worked by trial and error.[55]

The resolution of these contradictions was to declare Edison "the wizard of Menlo Park," who became the protagonist in many stories. Newspapers seriously reported that he was attempting to communicate with the dead, or that he had invented a paper shirt of many layers that could be peeled off, one by one, when they became soiled. On April Fool's Day, 1878, the *New York World* reported that Edison had invented a machine that could manufacture "biscuits, meat, vegetables, and wine out of air, water, and common earth."[56] Some papers ran that story as serious news. Yet other Edison stories focused on money and success. In one, he was the poor young boy from the provinces who climbed the ladder of success. Edison left home in his teens to become a wandering telegraph operator. Nearly penniless, the story went, he arrived in New York City on Black Friday, when the stock market was in a frenzy over the price of gold. Trading became so hectic that the recording machinery broke down. Messengers mobbed the office, trying to learn the correct price of gold, but no one was able to get the Gold Stock indicator to work. When Edison made swift adjustments to the indicator and it began to function, a grateful manager offered him a well-paid position. This story became a part of Edison's biography,[57] but in fact, Black Friday occurred months before Edison arrived in New York.

Other tales emphasized Edison's association with unscrupulous robber barons interested only in money. During the 1870s Edison worked for the notorious Jay Gould, who manipulated the value of telegraph companies and railroads with little interest in public welfare. For Gould, Edison invented a way to send four messages over one telegraph wire simultaneously. Edison's later development of the incandescent electric light was financed by J. P. Morgan. Such associations suggested Edison was not a worthy young chap from the provinces but a capitalist collaborator. His announcements about electric lighting drove down gas stocks and attracted investors to his projects. The *Saturday Review* suspected Edison profited from the rising and falling belief in his imminent invention of a practical electric light. It suggested that "there is no reason why he should not for the next twenty years completely solve the problem of the electric light twice a year."[58]

In some early Edison stories, his inventions were not a commercial success. One writer declared in 1884, "The failure of the phonograph did much to destroy the popularity of Mr. Edison. So did his subsequent failure to utilize the electric light in private dwellings to take the place of gas. So did his connection with the telephone, the honors of which discovery were carried off by his competitors."[59] That story did not long survive, however. The most common description was of a brilliant inventor who set up companies to manufacture his creations. His West Orange laboratories were eventually surrounded by factory buildings. His lab, or "invention factory," was arguably his greatest innovation.

In later life Edison helped four biographers to fashion a story that combined the story of a Horatio Alger boy who rises from poverty to success with an inventive wizard and a canny entrepreneur. This narrative was worked out in detail in two volumes published in 1929 by his close associate William H. Meadowcroft.[60] The "wizard of Menlo Park" united central contradictions of his age. He was the boy with little education who became a world-famous inventor. He represented both science and tinkering, method and magic, corporate capitalism and individualism, the rural world he came from and the high-energy city he enhanced. He believed in the infinite possibilities and benefits of inventions. In 1896, he declared during an interview that "every discovery cheapens articles and tends to bring everything within the reach of common people. . . . There is an endless scope for discovery and invention."[61]

Ford, Carnegie, Hoover, Crusoe, and Edison were characteristic heroes of the high-energy lifeworld. In the Crusoe story, one man completely defined and controlled the space around him, developing a desert island into a habitable, productive home. His story could appeal to someone homesteading or settling in a new place. It defined nature as a resource waiting to be exploited. The simplified Edison story said an inventor could create useful objects, establish new industries, benefit the public, and ensure progress. Edison also exemplified the small-town boy triumphing in the metropolis. In these narratives the individual flourished in a free market, with little need for government. Like Edison, Carnegie was a poor, largely self-educated boy who rose to the pinnacle of fame and fortune. If Edison had benefited the public through his inventions, Carnegie did so through philanthropy. In this ideology, individualism maximized each man's opportunities, ensured economic progress, and encouraged charity. The fabulous success of Edison or Carnegie was rare, but there were other examples, notably, Henry Ford, who lacked a high school education but became the world's largest manufacturer of automobiles. Likewise, Edward Bok came to America from Holland as a boy and rose to immense success as the publisher of one of the nation's largest magazines, the *Ladies Home Journal*. Bok's autobiography won the Pulitzer Prize in 1920 and went through many editions.[62]

While the success of Franklin, Edison, Carnegie, Ford, Hoover, and Bok might seem to validate the ideology of individualism, the contradictory stories about Edison are a reminder that the story of triumphant individualism was intertwined with narratives of ruthless robber barons, wizardry, failure, accidents, labor unrest, and injustice. The counter-narratives called for social programs to relieve the poor, for higher wages, for better working conditions, and for an end to monopolies. The ideology of individualism demanded that the nation be an even playing field, but class differences were widening, and more than half the population—women, unskilled workers,

Native Americans, African Americans, Latin Americans, and "undesirable" immigrants from China, Japan, and many other nations—were excluded or marginalized. Even for white men, the probabilities of success depended on factors mostly beyond their control. Tocqueville's concept of individualism assumed an egalitarian society with a minimalist government, where every person was the master of their fate. By 1929 that was seldom a description of actual conditions. Nevertheless, the enormous influx of oil, gas, and electricity had made possible new forms of employment, shorter working hours, higher real wages, and access to new consumer goods. These changes provided a general sense that upward mobility remained possible in urban America.

Individualism persisted. It provided an ideology of success that fit with the idealization of the small town. It was a link to the nation's agrarian origins, and self-reliance seemed a central element in democracy. The ideal American hero was a white man of humble origins from a small town, often with little formal education, who became a success through hard work. Many seemed to fit this description, including Edison, Ford, Carnegie, Bok, Hoover, and to some degree the aviator Charles Lindbergh. They invented new devices and founded corporations that radically improved the life-world. Because of them, people could record sound, own automobiles, enjoy libraries, live in electrified houses, and soar above the earth. Most of them were closely identified with the high-energy economy. Their fame and fortune arose in new industries based on oil, gas, mining, electricity, automobiles, steel, publishing, and aviation. The scale of their success took place within, and required, a national market.

III

MOMENTUM

8

NATIONAL MARKET

The great energy transition had made possible higher wages; shorter working hours; brilliant illumination; networks of water, gas, sewage, and communication; new forms of entertainment; and innovative consumer goods that transformed daily life. Americans in 1925 had more leisure time than their grandparents did in 1875, and they had an automobile instead of horses, an indoor bathroom with hot and cold running water instead of an outdoor privy, new electric appliances, and a larger wardrobe. They had seen things that did not exist in 1875: including phonographs, electric streetcars, motion pictures, automobiles, airplanes, X-rays, assembly lines, and radio. The control of new energy sources had made these things possible, but it had also facilitated the growth of new corporations that dominated the economy.

For the great energy transition changed the hierarchy of the nation's businesses. In 1876, the dominant corporations were driven by steam. The Pennsylvania Railroad was the nation's largest corporation, and at its peak in 1920 it employed more than 250,000 people. The New York Central Railroad was also enormous, and in all parts of the nation railroads were essential to local economies. To meet railroad demand for rails and rolling stock, steel production skyrocketed from 68,000 tons a year in 1870 to 4.2 million tons in 1890, and it continued to increase due to widespread adoption of steel beam construction. Yet the dominance of steam and steel began to wane after 1900, as could be seen in the makeup of the Dow Jones Industrial Average. When invented in 1884, the Dow was composed entirely of railroads, except for Western Union telegraph.[1] At that time the oil and electrical industries were still comparatively small, in stage two of their development in the Hughes model of energy transitions discussed in chapter 2. After a decade, the Dow was revamped and based on twelve corporations that relied on railroads to transport commodities and raw materials, including Amalgamated Copper, American Smelting & Refining, American Sugar, Continental Tobacco, International Paper, National Lead, and Peoples Gas. All but one of these twelve firms extracted or refined natural resources.[2] The exception was General Electric, whose sales were an indicator of the demand for electric ovens, dryers, pumps, mixers, drills, cranes, elevators, presses, streetcars, locomotives, and other machinery. Electrical equipment was being adopted in every conceivable industry. By this time the electrical industry had reached the technological momentum of phase four. Westinghouse, the other major electrical machinery maker, soon also became part of the Dow, along with the oil corporations and other industries that gradually replaced the railroads and extraction industries. By September 1929, the Dow industrial average was based on thirty companies, only three of which had been on the list in 1900. More than half had not even existed in 1876. Nine represented the growing demand for oil and the machines that used it: automobiles, trucks, airplanes, and harvesting machines. Three made electrical equipment. Six sold consumer goods, including General Foods, Woolworth's, and Sears, Roebuck.[3] The changed makeup of the Dow recorded the shift from resource extraction and steam power to electricity, oil, and high-energy consumption. The railroads had laid track and built magnificent stations. The high-energy corporations built skyscrapers, department stores, mail-order empires, oil refineries, power plants, and enormous factories.

These high-energy corporations were almost all run by salaried managers, many of them graduates of the new business schools. Compared with private companies, corporations had many advantages. Private owners were personally responsible when a firm went bankrupt, and they could lose their house and all their possessions. In contrast, stockholders in a corporation were liable only for their direct investments. Furthermore, incorporating a company made it easier to raise funds through the sale of stock, in order to expand operations or to purchase new equipment. And where a family firm might decline during the dotage of a founder or under control of an incompetent heir, a corporation could dismiss ineffective leaders and hire replacements. However, these administrative advantages alone do not explain why American corporations grew rapidly after 1870. They grew because the United States was a large market with no trade barriers, because the canal and railroad network provided inexpensive access to that market, and because the great energy transition provided unprecedented access to flexible heat, power, and light. By 1929 the new high-energy corporations had matured and dominated the American economy.

Their dominance was predicated on changes that made the United States a thoroughly integrated, national market. The Civil War had abolished regionalism in politics, but full economic integration had not yet been achieved. The canals, railroads, and telegraph linked the nation into a single market for raw materials such as lumber, iron ore, coal, cotton, corn, sugar, and wheat, but many consumer products were still local, produced in small quantities for regional markets. A national market also required a standardized national currency, brand names, and national advertising.

A powerful national economy had global consequences. In 1876 the United States was a successful economy, but it trailed behind Great Britain. By 1929, Americans had the world's largest economy, with the most productive workers and the highest standard of living, with one automobile for every 4.6 people. By comparison, there was only one car for every one hundred Germans.[4] The American dollar had become an international currency, backed by gold reserves that amounted to almost 38 percent of the world's bullion. In the 1876 the United States was a debtor nation, but in 1929 it was a creditor nation, and New York had become the financial center not only of the United States but of the entire world.

This was quite a change from the situation before the Civil War. After 1836 there was no national bank to guarantee the value of paper money, and local banks issued their own notes. These regional currencies were not always accepted at face value the farther one traveled from the bank that issued the notes. The public preferred silver or gold coins, but these were not entirely uniform. Made of soft metals, they soon became worn and uneven. Moreover, they were often in short supply. Before 1860, "The Mint rarely coined the smaller denomination silver coins—half dimes, dimes, and quarters—needed for daily transactions."[5] The public therefore used some foreign coins, which were legal tender until 1857. Many preferred Spanish gold coins. When coins were not available, barter was often resorted to rather than bank notes. During the Civil War, the Confederacy issued its own currency, and in 1861 the federal government began to issue "greenbacks" printed with "fine-line engraving, intricate geometric lathe work patterns, a Treasury seal, and engraved signatures to aid in counterfeit deterrence."[6] The dollar's stable value and widespread availability facilitated financial transactions, whether purchasing stocks via the telegraph or buying groceries in person.

In a national market, people had to be convinced to buy products based on their reputation rather than on the good repute of a neighborhood retailer. In local markets, one was a customer with a personal relationship to the clerk behind the counter. In a national mass market, one was an anonymous consumer. In the 1870s there was still some haggling and barter, but it was being replaced by fixed prices and cash purchases for standardized goods. The history of Woolworth's illustrates the emergence of mass marketing. Starting in 1879, it spread throughout the country from its first store in Lancaster, Pennsylvania. At first, all items were sold for five cents, but soon ten-cent items were added. Goods were sold with only a small markup from the wholesale price, with profits based on rapid turnover. Forty years after these beginnings, Woolworth's had grown into a chain of 1,200 stores, and the skyscraper headquarters in the Woolworth Building was so intensively lighted that on a clear night it could be seen fifty miles away. Woolworth's exemplified how mass production made possible mass consumption. All Americans shopped in its stores at least occasionally, and such chains of stores became characteristic in consumer society (figure 8.1).

Woolworth's was a brand name, and every company could see the value of establishing a brand of its own. Silversmiths and other skilled workers had put a small hallmark on the side or the bottom of their productions, and it is still used by experts to establish the provenance of antiques and luxury goods. Trademarks for mass-produced goods began to be protected by federal law in 1881, precisely when corporations were expanding into the national market.[7] Trademark protection was a necessity before a firm adopted national advertising, distinctive packaging, and brand names. Trademarks were assurances of uniformity, and they often became the corporate logo, sometimes in the form of elegant lettering (as with General Electric and Ford) but more often as a fusion of lettering and visual imagery.[8] Trademarks were well-suited for use in electric signs on the White Ways of American cities. By 1910 more than twenty blocks of Broadway were covered with a pulsating landscape of trademarks and slogans.[9] As an advertising handbook for salespeople explained, "The story is told in detail in magazines and newspapers, and the spectacular electric sign follows with its flashing presentation of the trademark."[10] Spectacular signs literally kept a brand name before the public (figure 8.2).

Trademarks were jealously guarded because infringement jeopardized a product's identity and interfered with customer loyalty. Often the trademark was not only on a package but stamped on the surface of the product, including Nabisco biscuits and the Louisville Slugger baseball bat, which

Figure 8.1
John Taylor Arms (1887–1953), *An American Cathedral (The Woolworth Building)*, 1921, etching. Courtesy of the Amon Carter Museum of American Art, Fort Worth, Texas.

Figure 8.2
Broadway and 45th Street, 1923, New York Edison Co. Photographic Bureau. Gelatin silver print. Courtesy of the J. Paul Getty Museum Collection.

became a registered trademark in 1894. A process of amalgamation began in which the trademark began to fuse with the object. The Coca-Cola bottle was advertisement, trademark, and product rolled into one. Packaging became part of marketing, including endorsements. In 1905 the Louisville Slugger company paid Honus Wagner, the star hitter of the Pittsburgh Pirates, for his endorsement. He was the first athlete in any sport to endorse equipment, and his signature was engraved on the surface of each bat.[11] In a national market, it made economic sense to link products with celebrities.

Endorsements and trademarks were part of a shift from products known through direct physical contact to products known through a symbolic shorthand that identified them as desirable objects. Theodore Dreiser was sensitive to this change in *Sister Carrie*, which recounts a young woman's awakening to the commercial culture of Chicago. After arriving from a small town, Carrie learns to adjust her dress, her gait, and her personality in accord with the latest urban styles. When shopping, she acquires more than garments. She studies the other women, learning how to wear the new clothing and how to carry herself. Taking cues from others, she redefines herself by adopting the changing fashions. "Every hour the kaleidoscope of human affairs threw a new luster upon something, and therewith it became for her the desired—the all. Another shift of the box, and some other had become the beautiful, the perfect."[12] For Carrie, consumer goods redefined and improved the self.

Corporations addressed the public by advertising in large-circulation magazines that emerged by the 1890s. Before then, national advertising scarcely existed. In the 1870s it was common only for the sale of books, journals, and patent medicines. With limited revenue from advertising, magazines such as *Scribner's* and the *Atlantic Monthly* had a high subscription price and addressed a literate elite. Compared with the mass-market publications that flourished after 1890, literary magazines had few illustrations and confined advertising to the beginning and the end of each issue. National advertising by corporations appeared throughout the pages of the new illustrated magazines such as *Munsey's*, *McClure's*, the *Ladies Home Journal*, and the *Saturday Evening Post*. They were larger and yet less expensive than the older literary magazines because of their advertising revenue. In 1898 Nabisco launched the first $1 million advertising campaign and focused its spending on these new magazines. Other large advertisers included the American Tobacco Company, Ford, General Electric, General Motors, Campbell's Soups, Heinz, and Quaker Oats. When readers flipped through the pages of the new magazines, they were in effect turning a kaleidoscope. The goods were often inserted into attractive sites, associating the product with national parks, skyscraper cities, or pastoral scenes. Readers saw both American scenery and a cornucopia of products that together offered a new vision of the national. The local faded by comparison.

Moreover, people had many more years of life to make purchases and to expand their horizons. Life expectancy rose from forty-four years in 1870 to sixty in 1930. That meant sixteen more years to work, to consume, and to travel. The population's median age increased from 20.2 to 26.5 years.[13] Wages increased modestly in any one year, but during a lifetime they skyrocketed. Ordinary Americans born after 1870 experienced dramatic property mobility, as they acquired many objects that were rare or did not even exist when they were growing up, including indoor plumbing, central heating, electric lights, phonographs, and automobiles.

However, prosperity and longevity were unevenly distributed. The wealthy lived longer than the poor. Life expectancy for African Americans was only thirty-three years in 1900, rising to forty-five by 1930. And they had lower incomes.[14] New goods were more readily purchased by the white middle class, which had higher incomes and lived longer. Their consumption increased at the same time as the workday in offices and factories gradually declined by three hours. The historian Thomas Schlereth succinctly explained the result: "More people (middle class and working class) had more money and more time to purchase more goods, mass-produced more cheaply and advertised more widely."[15]

In this emerging lifeworld, products were no longer expected to last a lifetime, or for even a decade. They were in constant evolution, becoming, advertisements proclaimed, "new and improved." Henry Ford resisted this trend, but it was embraced by General Motors and other car manufacturers. Ford's Model T retained its appearance for fifteen years. There were improvements of individual parts, but when put together in 1927 they resembled the car from 1914. In contrast, General Motors rolled out new models every year. Their changing appearance registered the growing importance of styling and design. The consumer selected goods less on functional criteria than as a matter of taste. New styles, colors, and visual vocabularies such as streamlining were introduced.[16] As industrial design matured, Miles Orvell found that successful products laid claim to a heightened reality. The public wanted to own these "real" products as accoutrements of a better self. They became obsessed with authenticity, determined to possess "the real thing."[17]

Many advertisements reversed the defamiliarization that was a common strategy in early twentieth-century modernist art. Through defamiliarization an artist asked the public to see the world afresh by "making strange" an everyday object that had become so familiar that it was encrusted with banality. The underlying idea was that the organs of perception could be "cleansed of the mystifying and misleading accretions which result from our experience of a corrupt and corrupting world."[18] The modernist artist wanted to "make it new." In photography, for example, one could make the familiar strange by using odd angles, mirrors, extreme close-ups, distorted shadows, juxtapositions, and collage. While de-familiarization was a central idea in modernism, however, corporations often sought just the opposite. They wished to make new products seem familiar. General Electric designed its first appliances and lighting fixtures so that they fit seamlessly into a home. They were presented not as cutting-edge technologies that would link the house to an electrical network, but as improved versions of familiar devices.[19] For advertisers, a common method of doing this was to insert a new product into what Marchand called a social tableau, a somewhat idealized "slice of life" that was "sufficiently stereotypical to bring immediate audience recognition." These tableaux were less mirrors of society than mirrors of aspirations. They usually depicted "settings at least 'a step up' from the social circumstances of the readers."[20] The new products were ensconced in a familiar household interior. They both blended in and yet stood out as desirable additions. As Lears explained, "No matter how fervently they chanted the gospel of newness, advertisers knew they had to establish some common ground, some sense of old-shoe familiarity between the purchaser and the product."[21] Successful consumer goods, whether a radio or a toaster, ought to fit seamlessly into a familiar tableau, as General Electric understood when it issued a brochure entitled "Home of a Hundred Comforts" (figure 8.3).

Figure 8.3
"Home of a Hundred Comforts," 1925. General Electric. Courtesy of the Hathi Trust.

Americans were encouraged to express their identity through ownership of consumer goods. As Marchand demonstrates in *Advertising the American Dream*, the insistent messages of mass culture fostered the concept that most people belonged to a vast middle class. Advertisers knew that there was a wealthy segment of 5 percent of the population, whom they addressed differently. But most advertisements were aimed at the middle class, with the assumption that the poor aspired to the same goods.[22] When Black people appeared in advertising, they were depicted as servants, cooks, gardeners, and the like. Mass culture was conceived and addressed as being white.

Advertising executives were mostly college-educated men (and a few women), who regarded the middle class as "intellectually inferior, culturally backward, emotional, and tasteless" but capable of improvement.[23] Advertisers sought to upgrade popular taste but repeatedly discovered that logical arguments were less effective than emotional appeals. Accordingly, the space used for text declined in favor of images, and statistical or logical arguments gave way to anecdotes and stories. The voices of advertising became casual, intimate, even frivolous. Marchand notes, "Just as the newspaper reader chose the tabloid picture over the serious story and the comic page over the foreign news, and just as the radio audience turned its dial to tasteless comedy rather than classical music, so the consumer would not accept serious advice about products in any medium without a dream world of frivolity and fantasy to go with it."[24] The task of advertising was to construct tableaux that inserted products into an idealized middle-class life.[25] This fed into the consumer imaginary of a lifeworld of continual improvement.

Advertisers depicted not the consumer's class position but their aspiration to climb the social ladder. They showcased domestic interiors illuminated by the new electric lights and that had phonographs, telephones, electric fans, toasters, waffle irons, hot plates, refrigerators, stoves, and radios, all of which entered American homes after 1890. These aspirational products were both time-saving appliances and the latest fashions. Women were encouraged to imagine themselves as homemakers married to businessmen.[26] In these new domestic landscapes, technologies were "silent servants" that would reduce the labor of cooking, cleaning, or doing the laundry, and make more time available to listen to the radio, play the phonograph, read to the children, go to the movies, or chat on the telephone. In practice, however, new domestic technologies had unintended effects. They were used to transfer labor from husbands and children to housewives.

Housewives were expected to clean more often, prepare more varied meals, and do laundry several times a week. As Ruth Schwartz Cowan's work demonstrates, larger homes, a greater variety of available foods, and closets full of mass-produced clothing all meant more cooking, cleaning, and laundry for mother.[27]

The new factory jobs, offices, department stores, and electrical networks were in cities and suburbs, and consumption focused there as well. Most advertising campaigns were placed in periodicals aimed at urban audiences. Their collective circulation was more than eight million.[28] In contrast, most of rural America was not yet electrified, and appliances could not be plugged in. Some devices, like the early phonograph, could be wound up, but its volume could not be amplified. Farmers could listen to battery-operated radios, but they received fewer stations than their cousins in town. Batteries ran down quickly if used to run vacuum cleaners, irons, or toasters, and such appliances were impractical for those living beyond the grid. Different advertising addressed the rural population in magazines such as *Country Life* and *Keith's Magazine for Home Builders*. Taken together, their circulation was below one million.[29] However, the new goods were available through mail order. The extent of this business is exemplified by the history of Sears, Roebuck and Company. Twenty years after its beginning in 1886, its annual sales were $53 million.[30] The 1902 catalogue was more than 1,000 pages long, and it stimulated dreams of comfort and luxury for rural dwellers. It sold everything from guitars and zithers to mailboxes and ice boxes to the majority of households that did not have electricity. More than 600,000 copies of the Sears catalogue were distributed in that year alone. Rural America had access to a cornucopia of mass-produced goods, including all sorts of clothing, blacksmith tools, "talking machines," bicycles ($8.95 to $15.75), a candle or gas chandelier ($3.60), a black enameled iron trunk weighing fifty-one pounds ($4.69), or a cast iron kitchen range that could burn coal or wood and weighed 415 pounds ($14.95, plus $2.00 shipping for up to 500 miles).[31] Mail-order sales were further assisted by the introduction of parcel post in 1913.

The mergers into large corporations, the creation of brand names, the standardization of products, the development of annual models, and the wide distribution of products occurred along with the development of anonymous shopping. Markets and cafes might be closed, but consumers wanted food and cigarettes at all hours of the day and night. To meet this demand, after 1895 vending machines became common, making some goods perpetually available. These machines became a practical possibility once coins were plentiful and goods were branded and standardized.[32] The stronger the trust in a brand, the easier it was to sell products from an impersonal vending machine.[33]

Vending machines, including telephone boxes, were placed in public spaces, such as a hotel lobby, a railway waiting room, a gasoline station, or the hallway in a public building. The manager of a novelty company in Chicago observed in 1902 that in 1890 "there were less than two hundred coin-operated machines in the United States, most of which had been imported from Germany" as well as some produced in "two small machine shops, one in Chicago and the other in New York." They made devices that told a person their weight, dispensed chewing gum, or sold postage stamps. A decade later twenty companies manufactured machines that sold a "wide variety of articles, such as salted peanuts, gum, chocolate, candy, soap, postage stamps, hotel stationary, shoestrings, medicines, cigars and cigarettes." One company sold more than a million pounds of peanuts through 6,300 machines in four months. There were also machines that told one's fortune or provided a mild electric shock that was "therapeutic."[34]

Another early vending machine was Thomas Edison's kinetoscope, invented in 1891. It projected a short loop of film on a white background for one viewer. Edison thought it would be useful as an educational device or to demonstrate products, but its chief use was entertainment. In 1895, kinetoscopes showing short silent films of boxing, acrobatic feats, and other activities netted Edison $150,000 a year.[35] Other entrepreneurs offered piquant scenes, such as couples kissing, a scantily clad dancer, or other mildly erotic material. These early films were brief and had little or no narrative, and most kinetoscopes catered primarily to men and boys.[36] They were placed in a new penny arcade,[37] a room full of vending machines that offered short films, recordings of music and monologues, peanuts, postcards, and chewing gum, as well as "punching bags to compare your strength" with Corbett, Jeffries, and other boxers.[38] A Detroit firm "developed a large number of lung-testing, combined lung and grip-testing, punching, lifting machines," including the "Mickey Finn" that was a "combined tug-of-war and grip machine."[39]

The development of vending machines parallels changes in consumption that can be illustrated by automobiles, children's toys, and amusement parks. The consumer's early relation to all of these was kinetic. Certain physical actions, such as pulling a lever, had a cause-and-effect relation to getting a product from inside the machine into one's hands. For example, early machines that sold postage stamps required the consumer to put a coin into a metal arm of the sort still used in laundromats. Sliding in the metal arm pushed a coin inside, and pulling it out again turned a gear that unrolled the stamps. Variations of such machines were developed to provide candy bars, soda pop, and cigarettes. If the vending machine jammed and did not deliver the goods, the consumer customarily shook it, hit it, or kicked it. This was a logical extension of the kinetic, physical relationship that one had with the machine. The consumer knew that if it malfunctioned, a well-aimed slap or blow might release a hidden spring or jar loose a gear and un-jam the mechanism.

Similarly, mechanical toys for children often had a spring inside that was wound up by a crank or a key. If a child pried open a wind-up truck, visible inside were springs, gears, and other mechanical parts, and one could see how the toy worked. This transparency was part of its appeal.[40] Just as the child could learn how the mechanical toy functioned, early automobile owners who looked under the hood could understand the relations between the parts and see how to make repairs. The first generation of vending machines were part of a lifeworld with a direct physical linkage between action and outcome, like the child who could make a toy work by winding it up. Whether a child's toy, a vending machine, or an automobile, there were clear cause-and-effect connections. This kinetic relationship was also expressed at amusement parks. The typical rides were intensely physical, heaving patrons into the air on the Ferris wheel, knocking them down in the rolling barrel, or hurling them through space on the roller coaster. In contrast, later amusement parks, exemplified by Disneyland, offered patrons less kinetic and more visual experiences of landscapes such as the African jungle, the Wild West, or outer space.

The automat was another mechanical consumption system. Its walls were covered with small glass compartments. Behind every glass door was a sandwich, piece of cake, or other food item, and each cost a nickel. Such automated cafes emerged first in Germany and France, and one was popular at the 1897 Brussels world fair.[41] In 1902, Horn and Hardart opened the first American automat in Philadelphia. Such informal and inexpensive dining appealed to office workers and shop assistants on short lunch breaks and to shoppers in a hurry. The automat proved especially popular in New York City, where millions of customers inserted their nickels.

As consumers became anonymous, they sought authenticity. As Orvell concluded: "It was as if there were some defect in everyday reality that had to be remedied by the more authentic reality of the object to be consumed."[42] Standardized products such as a GE radio or Ford Model A could be extremely desirable. The neo-Marxist critique of capitalism held that advances in production demanded national advertising to sell the flood of mass-produced goods.[43] In this view, corporations created and controlled the mass market through advertising, industrial design, marketing, and public relations.[44] But this argument assumed that consumers were passive and easily manipulated, and it paid little attention to the new goods themselves.[45] It was not so simple. As Fredric Jameson realized, products embraced by popular culture often served a utopian function. The consumer was not passively imbibing ideology but actively seeking pleasure and release. Consumer culture could contain subversive longings, and popular amusements could cater to yearnings for a different life.[46] Films, music, novels, and other products of popular culture did not necessarily impose a false consciousness on the consumer. Rather, they afforded ordinary people an escape from routine into fantasies of another kind of existence. Even implausible fantasies expressed dissatisfaction.

William Leach applied this insight to the department store and found that as early as 1900 it offered female shoppers release from domestic confinements, parsimonious habits, and Victorian codes of behavior. Goods were displayed as parts of larger tableaux, creating a fantastic interior, at times embellished by exhibitions of modern art. Before 1920 the Gimbels department store displayed paintings by Cézanne, Picasso, and Braque; in Chicago, Carson's hung canvases of the Ash Can School in its store gallery. The department store was a high-energy palace of consumption, making shopping pleasurable and theatrical, as the well-dressed crowds gazed at the chandeliers, rode in the elevators, or looked at each other, measuring their position.[47] Shopping became a continual negotiation between the self and dazzling new styles and transformations. Consumption on the elaborate stage of the department store seemed to offer a cornucopia of possibilities and liberation from care and toil. In these years Thorstein Veblen coined the term "conspicuous consumption" to describe the symbolic uses that consumer goods had achieved, as the middle class learned not merely to display wealth, but to act out new social roles, using material goods as building blocks of identity. As the anthropologist Mary Douglas and Baron Isherwood later argued, "the most general objective of a consumer can only be to construct an intelligible universe with the goods he chooses."[48] Dreiser understood this when he described Carrie Meeber's shopping.[49] Corporations did not "colonize leisure"; rather, emulative spending and styling colonized production.

If consumer demand was central to the culture of consumption, what did consumers want? They preferred goods that were new yet familiar, innovative yet timeless. An object's appeal (usually) did not last very long, as the preferred style for objects kept changing. W. J. T. Mitchell goes far toward defining the contradiction within each consumer good, which is "in theory 'only a symbol' (thus not a fetish)," and yet being "in practice 'the thing itself' (thus a fetish)."[50] This contradiction is central to the life cycle of goods, which are subject to iconoclasm and demands for continual renewal. A product oscillates between the symbolic and the material until it seems obsolescent and is discarded. An object often loses its symbolic attraction long before it has lost its utility. Later, it may be resurrected as an emblem of a former era.

Consumer goods are not abstract. They have kinetic appeal. A Louisville Slugger baseball bat is not merely a pleasing form but an extension of the arms. It provides a physical experience to all who know how it feels to hit a ball in the "sweet spot." Anyone who has played baseball knows how it feels to heft and swing the bat and has an aural memory of its crack when hitting a ball. The consumer knows objects through the senses. The first automobile did not merely symbolize speed and mobility; it was used for exploration, self-gratification, and escape. It supplied more power to people than they had ever controlled before. It displayed personal taste, both through the manufacturer chosen and through added accessories. The psychological investment in the automobile was more than a flaunting of taste and wealth. Driving extended the senses and empowered the driver. The philosopher Don Ihde notes that technology redefines experience. The driver learns to sense a car's limits, for example, in taking a sharp turn or in parallel parking, so that "one's bodily sense is 'extended' to the parameters of the driver-car 'body.'"[51] Driving involves tacit knowledge that is dynamic and perceptual. This knowledge evolves as products are upgraded with new functions and features.

The automobile, unlike the durable goods a person inherited in the pre-industrial period, was a transient possession that seldom lasted much more than a decade. In the high-energy economy, few goods were handed down through the generations. The car owner used it to expand the body's parameters, but its rapid obsolescence underlines how unstable identity became when grounded on consumption. To lack a car after owning one involved a loss of kinetic contact and a shrinkage of bodily experience. A new car usually included new features that enhanced experience. Mastering the capabilities of complex products renewed a sense of mastery, placing the buyer inside the *tableaux vivant* of an idealized present.

Products became both symbolic and dynamic enhancements of experience, whose sensory pleasures illustrate how intensified access to energy made possible new forms of living. Products such as a Model T or an electrified kitchen suggested a potentially new existence. They were not foisted upon a passive public; they catered to desires. As Michael Schudson has argued, the supposed power of advertising has been overstated, not least by the industry itself.[52] By the 1880s manufacturers found that they could not dictate new colors or designs, and that they needed help to find out what the public wanted. Regina Blaszczyk's *Imagining Consumers* demonstrated that corporations relied on "fashion intermediaries" who discerned changes in consumer taste. There was a constant interaction between factory designers, salespeople, department store managers, retail buyers, materials suppliers, art directors, home economists, advertising executives, and market researchers. Before 1900 the "field letters of factory salesmen read like primitive market research reports."[53] In the 1920s the public forced Henry Ford to abandon first the color black for the Model T and then the Model T itself. Consumer desires crystallized around particular objects, styles, and colors, and lacking this information or ignoring it could be costly. A leading executive at the J. Walter Thomson advertising bureau declared that "the consuming public imposes its will on the business enterprise."[54]

Few products were as popular as the Model A Ford, introduced on December 2, 1927. Consumers literally lined up to buy it, waiting for hours or even overnight. In New York City crowds gathered during the night at the showrooms. By dawn, all the surrounding streets were jammed, and the police were called in to restrain the prospective customers. There were similar scenes in Detroit, Pittsburgh, New Orleans, and the rest of the country. The first day, 15,000 people made a down payment on the Model A without taking a test drive.[55] The new Ford became iconic the day of its introduction. To attain such a status, advertising was not enough, and durability was not enough and often not essential. Walter Benjamin wrote a famous article about how the unique work of art lost its aura in the age of mechanical reproduction.[56] Yet in practice a kind of aura has at times been attained by mass-produced objects. It is misguided to dismiss the powerful hold they have on the imagination as a mere product of clever marketing. Successful mass-produced objects fuse form, function, and kinetic appeal into a satisfying whole. They offer new bodily experiences. The automobile made possible a new sense of self, as did the elegant department store. Like Dreiser's Carrie Meeber, Americans sought products and spaces that enabled them to restage the self within an upgraded lifeworld.

The tableaux of an ever-changing culture of consumption were presented as scenes at world's fairs, in mass circulation magazines, in store windows, and inside urban emporiums. This imaginary was based on intensified energy use and mass production, stimulated by higher wages, enjoyed during increased leisure time, and constantly renewed through advertising and design. The great corporations played a central role not only in the production and distribution of goods but also in constructing a consumer imaginary.

The social and psychological effects of the corporate economy were complex. Every Woolworth's drove local stores out of business. The chain stores negotiated lower wholesale prices, pressured suppliers for discounts, and often held down wages. Other aspects of the new corporations were also disturbing. They could dominate a city, like General Electric in Schenectady, New York; Firestone Tires in Akron, Ohio; or the Ball Glass Corporation in Muncie, Indiana. Some companies and towns had the same name, notably, Pullman, Illinois; Hershey, Pennsylvania; or Corning, New York. Corporate employees remained nameless to upper management. Assembly-line factories subdivided work into jobs that could be learned in a matter of minutes and which in a week became a numbing routine. Yet because wages often rose while working hours fell, the fiercest criticism of corporations often focused elsewhere. They were blamed for commercializing culture through the excessive use of advertising, for standardizing goods, and for being undemocratic institutions with a hierarchical organization modeled on the military. They were accused of gaining unfair advantages in the marketplace through skillful use of lobbyists and tax lawyers. And they were criticized for being politically conservative institutions that seldom supported progressive policies.[57]

Another aspect of the corporation has received little attention, however, namely, how its identity splintered and multiplied in the early twentieth century. A corporation could be many things simultaneously, appearing as a good investment opportunity, a vast factory complex, a scientific institution, an extended family, a community benefactor, a guarantor of progress, or negatively as a capitalist institution bent on profits, a lobbyist focused on its own interests, a heartless exploiter of labor, or the purveyor of exaggerated advertising for unnecessary products. To some extent the contradictory images of the corporation expressed the different points of view of labor, reformers, consumers, and journalists. But to a considerable degree this dizzying multiplicity also was due to communication strategies adopted by the corporations themselves.

By the 1920s many of the largest corporations like DuPont and General Motors had decentralized. Rather than exercising tight control from the top, they discovered the advantages of setting production goals and then granting each division freedom in meeting them, so long as they maintained satisfactory profit margins. As an unintended side effect of this process, corporations began to develop multiple identities. Where the classic family firm was closely identified with one founder and one factory, the corporation was diffuse. It was not owned by a family but by stockholders. It had not one factory but many, located in different cities. It had multiple relationships with society, each of which required different forms of communication, including publications and advertising. For example, starting in 1920 Westinghouse Air Brake published a magazine aimed at its workforce "to promote the general well-being of the employee and to facilitate pleasant, harmonious industrial relations."[58] It addressed only workers, not engineers, customers, or managers, each of whom received different publications.

General Electric also established a complex internal communications system. It had a large photographic department, professional in-house artists, and advertising specialists. They created text and imagery for magazines aimed at different audiences and markets: engineers, workers, progressive reformers, managers, and consumers. Its publication for engineers, the *General Electric Review*, had a layout that resembled a scientific journal.[59] With a monthly circulation of about 6,000, it addressed the engineers at street railways, utilities, city governments, or corporations. Its technical articles were illustrated with equations, charts, and statistics to demonstrate that GE was a scientific institution. The articles did not emphasize profits or costs, but rather safety, service, efficiency, and product durability. Its photography was restricted to machines against a white background, such as an electrically operated oil switch; product installations, such as dynamos in a new power plant; portraits of notable large machines, such as giant turbines; and portraits of engineers wearing coats and ties. Workers were seldom depicted and never named. They were seen from a distance and usually served to illustrate the enormous height or width of a new machine. The imagery and layout were severely functional and bore no trace of modernist photographic technique. The magazine encouraged engineers to think of their work as the logical outcome of research and development. The dignified portraits of engineers associated them with management rather than with the unnamed workers standing on one another's shoulders to emphasize the size of a turbine. Publishing such a magazine was part of a largely successful effort to treat engineers as professionals allied with management rather than encouraging ties to workers. Many corporations published similar magazines.

If workers had no identity and little dignity in the pages of the *General Electric Review*, they were at the center of the *Works News*, a magazine that addressed them alone. Begun in 1917 during the company's Americanization program, it was given to blue-collar workers twice a month. During the 1920s its circulation was larger than that of the *New Republic* and the *Survey Graphic* combined, and it was printed on better paper and had more illustrations than either of them. It came in several editions, each addressing employees in a different factory. These magazines did not encourage workers to identify with or meet workers elsewhere in the corporation. Rather, the local editions reinforced their separation from one another. The *Works News* seldom discussed engineering or research, but emphasized individual workers, usually skilled men, who had challenging jobs such as a crane operator guiding heavy loads through a factory. Assembly lines were seldom depicted. Many articles were not about work. They presented the corporation as a community where workers contributed to pension plans, took evening courses, held dances, celebrated Christmas, got married, and went on vacations. Many articles were about sports in industrial leagues, and they celebrated teams that won trophies. Workers, nameless in the engineering magazine, gained an identity in the *Works News* through socializing and athletics.

In 1921, University of Chicago researchers surveyed the in-house publications for workers at 175 companies.[60] Two-thirds of them had more than 1,000 employees. Their magazines were intended to maintain good relations with a workforce so large that managers could not know them individually. The corporations portrayed themselves as large families. The magazines carried stories on how to avoid industrial accidents or how to reduce waste. They gave advice on saving and home ownership, and many companies offered free financial counseling. They encouraged courses in English because workers with poor language skills misunderstood instructions, could not communicate easily with others, and therefore were more accident-prone. Much space was devoted to Americanization, with articles on "the principles of our government; advice as to obtaining citizenship papers; and stories about opportunities and advantages of this country." International Harvester had a campaign "to make every employee an American citizen," and many magazines printed "the biography of some prominent American who came to this country as an immigrant."[61] These publications rarely referred to the economy, interest rates, or profits, and they did not discuss the company's annual report. They seldom mentioned unions or strikes. Such controversial matters were left to oral discussion with shop committees.[62] The purpose of these magazines was to improve morale and increase corporate loyalty.[63]

Worker magazines usually were part of welfare capitalist programs that companies developed as an alternative to unions. They offered a range of services and programs designed to make workers feel at home. There was bocce bowling, baseball, and basketball. There were garden plots, savings plans, suggestion boxes where workers could offer advice, and patriotic observances, such as fireworks on July 4. During the prosperous 1920s worker magazines were somewhat effective in preventing unionization. But they were less mirrors of a new corporate harmony than a managerial dream. After the stock market crash in 1929, this idealized workplace clearly did not exist, as many workers were fired or put on part-time employment. Their relationship to management could no longer be defined by sports, educational programs, Americanization, and gardening. Many worker magazines disappeared. GE discontinued the *Works News* in 1930 and brought back a shorter version the following year. Like its predecessor, the new magazine continued to emphasize individuals, while seldom showing workers collectively.

Another GE magazine, *Light*, dealt with commercial illumination, and addressed city engineers, lighting contractors, and others involved in the selection and installation of streetlights and commercial lighting. Already in 1905 there were 75,000 electric advertising signs in the United States, and in the next decade almost every city and town intensified its street lighting.[64] To meet this demand, GE built a complex in East Cleveland staffed with specialists. *Light* promoted the product line and displayed installations, with an emphasis on lighting as not merely a functional improvement but as a progressive reform. It defined GE as a part of the City Beautiful movement. Brilliant illumination could encourage citizens to enjoy the city at night, deter crime, reduce accidents, discourage littering, and stimulate civic pride. GE also promoted technologies that upgraded urban infrastructure: traffic lights, improved sanitation, smoke abatement, and illumination of playgrounds and parks.

Another GE magazine, *The Monogram*, addressed local managers throughout the company, providing a vision of how the various units were linked. As Alfred Chandler noted, when large industries internalized production that previously had been carried out in many different companies, management became more complex. "As the number of workers required for a given output declined, the number of managers needed to supervise these flows increased. Mass production factories became manager intensive."[65] *The Monogram* covered a wider range of topics than any other GE publication, but it seldom depicted workers and devoted little space to Americanization or welfare programs. It was a channel of communication between upper and middle management. Images of the factory floor typically depicted no workers at all, implicitly saying that management controlled the space of production and oversaw promotion, sales, and installation. *The Monogram* was the only GE magazine that focused beyond the company. It depicted skyscrapers, fairgrounds, power plants, and distant electrified sites, showing the worldwide dispersion of company products. While the purpose of a magazine for workers was to promote loyalty to just one unit of the corporation, the purpose of *The Monogram* was to overcome the separation between the company's many units and divisions.

Collectively, General Electric's magazines depicted not one organization, but at least four. In the *General Electric Review*, engineers were told that the company was a scientific establishment concerned with inventing and disseminating new technologies. The *Works News* told laborers that every GE plant was a family that embraced paternalistic welfare capitalism. *Light* told illumination engineers that GE was a progressive company that was expanding the electrical infrastructure to make the streets safer, the air cleaner, commerce livelier, business more profitable, and all aspects of life brighter. *The Monogram* gave managers a comprehensive view of corporate activities, while seldom mentioning or depicting either engineers or workers.

Most corporate publications also contained images and articles about founders and inventors. At GE, the iconic founding fathers were Thomas Edison and Elihu Thomson. Just as important were the first director of the research laboratories, Willis R. Whitney, and the most famous scientist in those laboratories, Charles Steinmetz. He was depicted as an idealized inventor who worked for the sheer pleasure of discovery. He was tirelessly productive, whether at home, on vacation, or in the lab. Preoccupied with creativity, in images he usually held a new device in his hands. Such inventors exemplified the work ethic and the dynamism of a free-market economy. Their inventions made the corporation possible and illustrated progress.

Reviewing the many roles that corporations had assumed by the 1920s, Marchand found that corporate leaders saw business as a profession, like law or education. Its leaders spoke of the need for a broad historical outlook, for stewardship, and for public service. The corporation increasingly was an "institution" endowed with longevity. Like the law court, hospital, or university, the corporation was not merely a business but "an entity viewed as exemplary for society." An executive of Standard Oil of Indiana declared, "It is not enough to advertise a product. The public ought to be acquainted with the honest and high character of the institution back of the product." Gerard Swope, the president of General Electric, insisted in 1927, "My conception of industry is not primarily for profit, but it is for service."[66]

Business had certainly changed. Corporations had been more regional than national in the antebellum period. Their rapid growth after the Civil War was not intended, but it came about because of other interconnected changes. The creation of a national currency facilitated business transactions across the country; the legal recognition of brand names made it easier to establish and safeguard a corporate identity; improvements in mass production made possible larger quantities of identical, trademarked goods; presenting these products to the public required national advertising; and these advertisements made possible a host of new, inexpensive, mass-circulation magazines. Collectively, these changes encouraged corporations to consolidate production in large factories that mass-produced items for the middle class. Corporate trademarks offered consumers new badges of identity, and their products made possible new experiences and sensations.

In contrast to their carefully defined and protected trademarks, corporations themselves had evolved different identities for each audience. They were a bundle of contradictions behind the banner of being an institution. They were both patriarchal families and anonymous employers, businesses that maximized profits and philanthropic institutions, research laboratories whose scientists occasionally received Nobel Prizes and monopolies bent on the control of patents, the workers' friend helping them to become citizens and homeowners, and yet often the enemy of unions. In legal theory, corporations had the rights of individuals, but in practice they had multiple faces and adopted diverse discourses. They were all pervasive, yet elusively kaleidoscopic.

On a national scale, the high-energy lifeworld of 1929 also asked each individual to play multiple roles. Each was born into a racial group with certain advantages or restrictions, each had a vote in the political process, each attended a church or synagogue practicing freedom of religion, each had certain regional traits, each identity was further shaped by a job or profession, and each made crucial choices in the marketplace. Yet the leading newspaper in Muncie, Indiana, proclaimed, "The American citizen's first importance to his country is no longer that of citizen but that of consumer. Consumption is a new necessity."[67] In the 1920s, sociologists Robert S. Lynd and Helen Merrell Lynd studied Muncie as a representative American community. Since 1890 wages had risen faster than the cost of living, and families had acquired new "necessities" that included an automobile, a furnace to heat the house, "running hot and cold water, modern sanitation, electric appliances ranging from toasters to washing machines, telephone, refrigeration, green vegetables and fresh fruit all year round, greater variety of clothing, silk hose and underwear, commercial pressing and cleaning of clothes," as well as elaborate cosmetics, manicuring, and professional hairdressing.[68] All of these purchases either directly or indirectly demanded higher energy use, and most of them were supplied by national corporations and sold under brand names, whether automobiles, appliances, facial creams, or bananas. Freely spending in the marketplace, rather than saving, voting, worshiping, volunteering, or being a helpful neighbor, had become the epitome of good citizenship. This new consumer identity measured the self in terms of possessions. The Lynds concluded that older markers of identity based on labor unions, ethnic background, and religious beliefs were losing their importance. Instead, they noted "a constantly closer relation between the solitary factor of financial status and one's social status. A leading citizen presented this matter in a nutshell to a member of the research staff in discussing the almost universal local custom of 'placing' newcomers in terms of where they live, how they live, the kind of car they drive, and similar externals. 'It's perfectly natural. You see, they know money, and they don't know you.'"[69]

Consumers also embraced the convenience of gas, telephone, and electrical networks whose wires and pipes came directly into the house. In theory, control of more energy might also have been used to develop shared systems for heating, laundry, or transportation. But Americans rejected district heating (despite its greater efficiency) and put a furnace in every house. They preferred washing machines in each home to relying on laundries. They wanted private cars rather than mass transportation. They preferred single-family houses to apartments. Collective options were sometimes attempted with success in Europe, but in the United States consumers wanted an individualistic lifeworld. They wanted personal ownership, even if this proved more expensive and required more energy than collective solutions. Yet Americans were willing to embrace the collective experiences that the mass media made possible, as detailed in the next chapter.

9

MEDIA EVENTS

After 1876, the regional differences that led to the Civil War began to be described not as insurmountable divisions but rather as a pleasing variety within an increasingly united country. The nation was visualized not as semi-autonomous regions but as a continental country containing many picturesque sites that were celebrated in guidebooks, postcards, advertising, and newspapers. The citizen was increasingly defined as a consumer and part of the audience for media events shared across the nation, first using the telegraph, and later via telephone, film, and radio. Daniel Dayan and Elihu Katz define a media event as a decisive break with routine that attracts a large audience. While they focus exclusively on television, the media event emerged and developed for well over a century. Simultaneous national experiences became an important feature of the emerging high-energy lifeworld. It was a live event, giving it an element of unpredictability, but at the same time it was planned. For example, a government might organize the dedication of a monument, a presidential inauguration, or a state funeral. A city might celebrate the opening of a new bridge, park, or reservoir. Media events were not initiated by the media themselves, and they did not take place in their studios. Nor were they covered in the same way as news. Each one unfolded and was followed in real time. There was a script, but it was impossible to anticipate or control every aspect of an event, and there were often deviations and improvisations. Reporters did not stand outside and comment as critics. Rather, they functioned as narrators who conveyed excitement and participation to the audience.

Media events began on the local and state level before they expanded to the national stage. Even before the telegraph had been invented, Americans had begun to create celebrations that prefigured the media event. When the Erie Canal was officially inaugurated in October 1825, organizers placed cannons a few miles apart all the way from Buffalo to Manhattan. A cannon was fired when the first boat entered the canal in Buffalo, and when its blast was heard a mile or so to the east, a second cannon was fired, with many successive batteries echoing hundreds of miles along the canal to Albany and on down the Hudson River to New York City. Cannon-fire communication took hours to convey a prearranged message from one end of the state to the other. It announced that a procession of boats had departed from Buffalo, carrying dignitaries from town to town, each of which staged a reception and celebration.[1] The *Utica Sentinel* noted that "the whole proceeding was emphatically the work of the citizens—some classes were particularly distinguished, and none felt the generous impulse more forcibly than the mechanics and military:—their zeal was unbounded." The effect of the event was that "political animosity has for a time been banished by the generous burst of popular enthusiasm; and a whole people have poured forth the overflowing of their gratitude and mingled in a general acclamation of joy."[2] Such feelings of reconciliation and unity are often evoked by media events. In the case of the Erie Canal, the speakers emphasized that it would be much faster and cheaper to ship goods by water than to haul them in wagons. Canal boats would carry plows, stoves, equipment, and immigrants to the West and bring back the products of forests, mines, and farms. The canal would stimulate growth along its banks and in the Great Lakes region. The celebration was regional, but it had national implications.

The invention of the telegraph facilitated such events, making it possible by the 1850s to share a celebration simultaneously throughout much of the nation. For example, in 1858 huge public demonstrations greeted the news that Cyrus Field's Atlantic cable had successfully transmitted messages from Britain.[3] During the Civil War, crowds gathered at telegraph offices awaiting battle news, and during national elections people stood for hours in front of newspaper and telegraph offices, waiting for results.[4]

The inauguration of the first transcontinental railroad in 1869 was a dramatic early example of a media event. Like Lindbergh's solo flight from New York to Paris[5] or the Apollo 11 landing on the moon, this was a technological conquest of space, understood at the time as a further unification of the country and as the creation of a new passage to India by linking the Atlantic and Pacific Oceans. The last spike holding the track in place was made of gold, and it was ceremoniously hammered into place at Promontory Point in Utah. The occasion was memorialized in a famous photograph, showing two steam engines facing one another on the track, surmounted and surrounded by workers who had built the railroad and by dignitaries who had come to take part.[6] (The image showed a white-only crowd as it excluded the Chinese workers who built the rail line from California through the western mountains.) This ceremony occurred at a remote desert location, but it could be shared in all parts of the country, because each stage of the proceedings was telegraphed to waiting crowds, from New York to San Francisco (figure 9.1).

However, this early media event was unable to keep on schedule. The train from California arrived a day in advance of the ceremony, as agreed, only to find that the vice president of the United States and his party, traveling from the east, were delayed. Their train had been held up at gunpoint by angry railroad workers who had not been paid for months. They blocked the tracks and demanded $200,000 in back wages. It took time to get the $50,000 in cash that the workers settled for as a first installment. The train of dignitaries then began to move but faced further delay because trestles along the line needed repairs after spring floods. The ceremonies had to wait.

Figure 9.1
East and West Shaking Hands, Andrew Russell, photographer, 1869. Courtesy of the Library of Congress, Prints and Photographs Division.

Sacramento and San Francisco learned of these delays too late to postpone their ceremonies. Thousands of people had poured into those cities to celebrate, and they went ahead even though the final rails had not been ceremoniously hammered into place. On May 8, their bands played, parades marched, speeches were given, and thousands celebrated. The event was less simultaneous than planned, but it did not stretch out over several weeks, as was the case with the Erie Canal celebrations. Two days later, the Union Pacific train arrived at Promontory Point to hold the national ceremony. Telegraph wires had been attached to the silver and gold spikes, and across the nation crowds waited to "hear" the hammer blows that would link the Atlantic and Pacific Coasts. In anticipation, church services were held, speeches made, and crowds assembled in New York, Boston, Buffalo, Chicago, Omaha, and other cities across the continent. The event's precise timing was uncertain, but telegraph operators announced at 2:40 p.m. Eastern time that the prayers were over, and that Governor Stanford of California had begun his oration. His remarks lacked a memorable line. In a shorter speech, chief engineer Grenville Dodge proclaimed, "This is the way to India." Moments later the final spikes were driven in, with each blow transmitted coast to coast by the telegraph. The media event had reached its climax.

Immediately, cannons were fired in San Francisco and New York. Railroad whistles shrieked and bells rang across the country, including that in Independence Hall in Philadelphia. Waiting crowds broke into cheers and people poured into the streets and began to celebrate. Marchers in Chicago formed one of the largest impromptu processions ever seen there, and some estimated it was seven miles long. The news likewise brought crowds into the streets of Boston, Buffalo, St. Louis, and other cities.[7] Local festivities gained in significance because people knew that from coast to coast others were also celebrating. The continental railroad was understood as a symbol of American expansion, as an iron band holding the nation together, as a passage to the Orient, as a stellar example of the conquest of space and shrinkage of time, and as a stimulus to trade. The celebration itself was a new form of collective experience.

Future media events would be better organized. Specialists would plan and orchestrate them. Speech writers would polish the remarks to be made. Through the telephone and radio, the public would be able to hear them and through film would be able to see them. But the elements of a media event were already visible in 1869. It had to seem to be of national importance.

Dedicating a new library or town hall could only be local, but the dedication of the Statue of Liberty or of the Eads Bridge across the Mississippi at St. Louis had national significance. To increase the stature of such ceremonies required dignitaries: political leaders, prominent businesspeople, architects, builders, and engineers. The speeches ought to include striking phrases that could be cited afterward. The occasion required news coverage by reporters and transmission of the proceedings, at first by telegraph and later by telephone, film, and radio. The event had to be publicized in advance, to prepare the media to cover the story and to permit coordination between national and local arrangements. This ensured that bells, whistles, and cannons were ready, that banners and flags would be flying, and that crowds had gathered. A media event also required dramatic action that represented its larger meaning. In 1869, that eagerly awaited act was the golden spike being hammered in. The organizers in 1869 also understood that photographers could memorialize the event for posterity. Of the many shots taken that day, the iconic image was Russell's photograph of the two steam engines meeting, surrounded by celebrants, with men leaning out from each engine and holding champagne. People remembered that image and the golden spike, not the ceremony's delay, the railroad's dubious financing, its bankruptcy, or the reorganization needed afterward.[8]

The infrastructure needed to stage media events improved during the last decades of the nineteenth century. Railway trackage doubled, telegraph transmission more than quadrupled its capacity, and after 1876 the newly invented telephone spread across the country. Film emerged in the 1890s, followed by broadcast radio in the 1920s. Using these technologies, the burgeoning ranks of electricians, filmmakers, radio announcers, and other specialists with jobs that had not existed in 1876 made it possible for the entire nation to share dramatic events, helping to forge a national consciousness. Disasters such as the Chicago Fire (1871), the Johnstown Flood (1889), or the San Francisco earthquake (1906) immediately became front-page news everywhere. In Dayan and Katz's early research, these were not defined as media events because they were unplanned. But it later became clear to them that disasters ought to be included.[9] A fire, flood, or earthquake compelled national interest. Disasters were covered in newspapers, photographs, film, and radio. Even when newsreels arrived days or weeks after events were reported, they were avidly viewed in theaters, notably after the San Francisco earthquake.

No one anticipated the invention and development of media events. Today, such uses of the telegraph seem obvious. Yet at first, Americans did not know what to do with it. Businesspeople did not immediately grasp the advantages of early access to telegraphed information. When the first line connected New York, Philadelphia, and Washington, DC, business was slow. Six months after it began, it averaged only 2,000 words a day and made only about $600 a week.[10] In 1845 Samuel Colt built a line from Coney Island to Battery Park that telegraphed news of incoming ships to subscribing merchants. The venture was successful, and he soon operated a similar service in Boston.[11] By 1848 the telegraph linked Chicago to New York, and grain traders with immediate access to information made fortunes trading wheat. The telegraph consolidated the supremacy of these two cities as centers of commerce. By the 1850s the Chicago Board of Trade and the New York Stock Exchange were national institutions that used the telegraph to coordinate buying and selling. William Cronon summarized: "The wider the telegraph's net became, the more it unified previously isolated economies. The result was a new market economy that had less to do with the soils or climate of a given locality than with the prices and information flows of the economy."[12]

Most traffic on the wires was commercial, and the telegraph never became common in the home. Few people mastered Morse code. If they needed to send a telegram, they went to one of the many offices. An operator sent the message, which was often delivered by a youngster on a bicycle. The telegraph was rapidly adapted to a wide range of other functions. It could be used in monitoring and control systems, such as fire alarm boxes, stable calls, doorbells, and burglar alarms. Each communicated a single message: Open the door! Fire! Someone's here! Thief! In wealthy homes, wired communication called the servants. After the terrible fire of 1871, Chicago installed 430 miles of wires to link alarm boxes to the city's fire stations.[13] Police also installed call boxes throughout major cities. Telegraph technology became common in factories, which adopted watch clocks and surveillance devices. By 1875 an electric burglar alarm could be installed that signaled the opening of doors or windows. Electric fire alarms were also available.[14] Electric monitoring systems tracked the movements of night watchmen. As with the telegraph, many uses of a new technology are not foreseen. Machines are socially constructed, and the combined choices of inventors, entrepreneurs, workers, and consumers produced unanticipated results.

The telegraph played a central role in transmitting the news. Before it appeared, every small town had a newspaper that combined local news with whatever it could glean from other papers, often carrying stories that were a month old. Before Morse's invention, if a story seemed particularly important, a newspaper might charter a steamboat or make other arrangements to speed news to the public.[15] The Associated Press grew out of the intense interest in news of the Mexican War. Newspapers found that it paid to make extraordinary efforts, and five New York papers decided that it paid even better to pool resources.[16] From this collaboration grew the Associated Press, which soon had reporters telegraphing European news from the first ships to reach Boston (and after 1849 from Halifax). More and more newspapers became members. Dan Schiller found that objectivity in the news was not a product of this technology, which "was superimposed on a news-gathering system that already placed a premium on apparent factual accuracy."[17]

By the late nineteenth century, Trachtenberg notes, "factories, railroads, and telegraph wires seemed the very engines of a democratic future."[18] The "universal communication" of the telegraph was celebrated as a force holding the nation together. The *American Telegraph Magazine* declared that "nearly all our vast and wide-spread populations [will be] bound together, not merely by political institutions but by a Telegraph and Lightning-like affinity of intelligence and sympathy, that renders us emphatically 'ONE PEOPLE' everywhere."[19] Likewise, *Scientific American* praised the telegraph for promoting the "kinship of humanity."[20] But this did not hinder the use of the telegraph in the Civil War. Arguably, its rapid reports of events, such as John Brown's raid on a federal armory, fanned the flames of conflict. The telegraph also proved useful for the conduct of war, allowing generals to keep in touch with troop movements and to deal more quickly with supply problems.

Hopes for the realization of democracy through communications technology later shifted to radio. Commentators declared that it would spread "mutual understanding to all sections of the country, unifying our thoughts, ideals, and purposes, making us a strong and well-knit people."[21] In its early years, "radio was portrayed as an autonomous force" that could "reshape, on its own terms, the patterns of American life."[22] The unveiling of television later generated similar rhetoric.[23] Each communication technology initially seemed utopian, but such claims usually presaged the creation of monopolies or oligopolies. After decades of cutthroat competition, Western Union emerged as the dominant telegraph service. In the telephone industry, the Bell interests predominated and eventually swallowed Western Union into American Telephone and Telegraph. Nationally, radio was also concentrated in a few corporations. In each case, government regulation was needed to ensure competition within communications industries[24] (figure 9.2).

These new forms of communication were essential to media events. Pushing a button could activate a distant display or open a fairground, and this became common after 1880. For example, President Grover Cleveland pushed a button to open the Chicago Exposition. He was at the site on that occasion, but often politicians pressed a button in Washington. Woodrow Wilson pressed a button in the White House to open the Woolworth Building in New York, to open the Panama Pacific Exposition in San Francisco, and to celebrate completion of the Panama Canal.[25] Such push-button

Figure 9.2
General Operating Department, Western Union Telegraph, 195 Broadway, New York, 1875. Gelatin silver print. Courtesy of the Library of Congress, Prints and Photographs Division.

participation emphasized how networked communications linked all parts of the nation.

The electrical infrastructure could also be used to address audiences in several places at the same time. On July 4, 1916, President Wilson addressed several crowds simultaneously in New York City in a "silent speech" that was spelled out in electric lights. His rolling text could be seen simultaneously in City Hall Park, Times Square, Columbus Circle, and the City College stadium. The speech concluded, "May the light of America grow brighter with each generation and blaze out pure and undefiled to all the world."[26] This form of silent broadcasting was short-lived, however, as radio provided a more convenient and powerful way to reach large audiences.

Radio was at first used like the telegraph for point-to-point communication, but after World War I it emerged as a broadcast medium. KDKA, the first radio station, began broadcasting in November 1920 in Pittsburgh. It aired music and local news, and the following year it carried a baseball game for the first time. Other local stations emerged in all parts of the country. The first president to have a radio in the White House, and the first to speak to the public on the radio, was Warren G. Harding. His speech was broadcast at the dedication of the Lincoln Memorial on June 14, 1922. Few heard him, however, as most families did not yet have a radio, and there were only about thirty radio stations. Not all of them carried the speech, and some had weak signals. Just one year later, however, there were 556 stations, and by 1927 there were 681. After then, there was little growth in the number of stations, but their signals increased in strength.[27] Sales of radios shot up from $60 million in 1922 to $843 million in 1929. By the middle of the decade, it was possible to reach more than twenty million Americans through a radio broadcast.

Radio greatly amplified a media event's potential. Calvin Coolidge embraced the new medium and took lessons on how to speak effectively on the air. He was one of the first to realize that the stump speaking style that worked well when addressing a large crowd without a microphone was ill-suited to the new medium. He learned to modulate his voice and to keep a proper distance from the microphone. Most politicians had difficulty making such adjustments, for they were used to shouting to an outdoor crowd, while moving around and making broad gestures. Graham McNamee, one of the best-known announcers of the day, praised Coolidge's radio speaking style: "his voice came over in a perfectly even flow all the time, with never a syllable or tone being lost."[28] Millions heard Coolidge's first annual message to Congress in December 1924. He delivered forty radio addresses during his presidency, usually before a live audience. While Coolidge had the reputation of being taciturn, by 1925 he was delivering thousands of words a month over the airwaves.[29] In the 1924 elections, there were three national candidates. All made radio presentations, but Coolidge did so most often.

A list of 1920s radio media events would include the funeral of Woodrow Wilson in February 1924, heard by several million listeners, and Coolidge's inauguration in March 1925, which attracted an estimated 22,800,000 listeners.[30] Most of the time, the airwaves carried music, humorous programs like Amos and Andy, or sports. In baseball, radio made stars like Babe Ruth into household names. The World Series, which had been broadcast in various regions since 1921, became an annual radio event beginning in 1927. That year, the new NBC and CBS networks both carried the games. Dayan and Katz argue that such sporting events form a category of media events that are contests. A notable early example was the boxing match between Jack Dempsey and Georges Carpentier, broadcast in July 1921. A major example of the media event as conquest came six years later, when Americans followed intensely Charles Lindbergh's solo flight in 1927 from New York to Paris.[31] Newspapers described preparations for the attempt, but radio focused on the story once he was in the air. There was a large prize for the first pilot to complete such a flight, and Lindbergh was risking his life. Anxiety for his safety and excitement over his possible achievement culminated in national celebrations after he landed in Paris and again when he returned to the United States. He was received by vast crowds in New York, St. Louis, and Washington, where Coolidge welcomed him and delivered a radio address in his honor.[32] Lindbergh seemed to combine rugged individualism with mastery of the machine.

A speech, a baseball game, a parade, or an official reception all follow a script, or a set of rules known in advance, and they could become media events. In contrast, a drama such as Orson Welles's 1938 radio play "War of the Worlds," about an invasion from Mars, terrified listeners, but it was not a media event. It was fictitious, was made in a studio, was not known in advance, and did not take place in real time.

Media events were not inherent in the means of communication themselves; they were constructed. The telegraph, telephone, film, and radio were each greeted by utopian expectations. Each seemed to promise increased harmony and mutual respect. Each was expected to make the public more educated. Each was conceived to be a guarantor of democracy. But these anticipated effects were not inherent in the technologies, all of which in the United States were controlled by private corporations. The government could have decided to own the telegraph and the telephone, just as they owned the postal service. Some nations did exactly that. Radio might have been operated as a public service or treated primarily as an educational resource. Instead, private enterprise developed these communication systems. This maximized speed of implementation, but it also built into each system a profit motive, encouraged monopoly formation, and defined communications primarily as a business. These cultural choices shaped the form and content of media events.

Media events often celebrated an individual, marking an achievement or an anniversary. Dayan and Katz classify these as coronations, such as a presidential inauguration. This form of media event could focus on a private citizen, and one notable example was the celebration of Thomas Edison in 1929 at "Light's Golden Jubilee." His inventive career stretched from his telegraph patents of the early 1870s to the phonograph and electric light at the end of that decade to a host of additional inventions that poured out of his laboratories. Edison's inventions were crucial parts of the great energy transition. Yet, he had no control over the electrical business after General Electric was formed in 1892. Therefore, during the last forty years of his life the public did not blame him for the problems, policies, or prices of local utilities. In contrast to Alexander Graham Bell, whose telephone monopoly was one of the most hated trusts in America, Edison had no financial connection to the electrical industry he founded. In his later years he improved the phonograph, set up a movie studio, and worked on other projects. He represented more than his inventions; he became the model inventor who lived at his laboratory for days or even weeks at a time.

In 1904, when Edison's incandescent light bulb was twenty-five years old, a banquet for five hundred guests was held at the Waldorf Astoria Hotel, with the tables arranged like a giant letter E. Thousands of lights were hung about the hall. Replicas of his inventions made of sugar were on display, and for dessert there were ices shaped like electric light bulbs. Congratulatory messages were read from President Roosevelt, Lord Kelvin, Carnegie, and many others. He was called "the genius, our comrade, who took this little crystal bulb in his Promethean hand, and with it helped to give the world a glorious new light."[33] From this banquet emerged the annual ritual of celebrating Edison's birthday. Each year a contingent of reporters found him at work and their articles had headlines such as "Edison's Birthday Just a Work Day" or "Edison, 73, Will Work All Day." He spoke to reporters about a wide range of topics, including crime, taxes, life after death, outer space, the immorality of the jazz age, and politics. He advocated Prohibition, condemned Bolshevism, supported the French occupation of the Ruhr, and attacked the Ku Klux Klan. As such topics took precedence over inventions, Edison became an oracle.[34]

These annual events were a prelude to the international ceremony on the fiftieth anniversary of the invention of the electric light.[35] A postage stamp was issued in Edison's honor to mark October 21, 1929, as Light's Golden Jubilee, probably the largest media event held before 1930. One of the first international radio audiences listened in as he was canonized. General Electric and the Ford Motor Company sponsored the Jubilee, assisted by a public relations expert, Edward L. Bernays. Media events had come a long way since the continental railroad's completion sixty years before. Every aspect of the ceremony was carefully scripted, leading up to the climax when Edison reenacted the invention of his light bulb, assisted by Francis Jehl, a member of the original research team from 1879. The recreation took place in a meticulously reconstructed Menlo Park laboratory. The event selected from the facts of Edison's career so that his life matched the dominant popular story about him. It ended with a banquet in a full-scale replica of Philadelphia's Independence Hall, with five hundred international dignitaries in attendance, including the president of the United States.

At 9:30 in the morning on October 21, 1929, President Hoover arrived at gate 10 of the Ford Motor Company in Dearborn, Michigan, and alighted from his executive railway car. On the platform waiting in the rain stood Henry Ford, Harvey Firestone, Edsel Ford, and the guest of honor, Thomas A. Edison, with their wives. Behind them, press photographers snapped images. The president smiled and with his hosts began to walk, trailing behind him twenty-seven reporters, fourteen photographers, eight Secret Service men, several members of his cabinet, and eight railway officials. This entourage boarded a much older train restored for the occasion. In Edison's youth, this wood-burning locomotive had carried passengers and produce between his home in Port Huron and Detroit. On this train, the inventor had once published a small newspaper, gleaning news from the telegraphers along the line and from the day's newspapers. The train was a direct link to the inventor's humble origins. Every day it had taken him to Detroit, where he frequented the public library and educated himself for a few hours before the afternoon return trip. On board, he had sold fruit, snacks, magazines, and newspapers. The train became a symbol of his life's journey from rural obscurity to the pinnacle of fame. Edison grasped this symbolism and picked up a basket of fruit and began to hawk his wares to the president and other passengers. With this gesture, Edison made explicit a central element of the narrative at the heart of the day's events: He had been self-reliant. That evening at the banquet Hoover told millions of radio listeners that "by his own genius and efforts [Edison] rose from modest beginnings to membership among the leaders of men, and his life gives inspiration and confidence to every youth in our land, that the door of opportunity still stands open."[36]

This was not the only symbolic link between Edison and the train. At Smith Creek Station he and all his chemicals had been thrown from the baggage car after some phosphorus accidentally caught fire. The toastmaster that night, General Electric executive Owen D. Young, interpreted that event as a crucial moment in his career. Certain his audience already knew the story, he only briefly alluded to it and concluded, "That was a great day for the electrical industry. One might say that it was largely born at Smith Creek's Station. The same unconquerable will and unquenchable fire which took you out of that depot, which set you out on your career, brings you back to it today." Young presented ignominious dismissal as the prelude to success. The return in triumph to a scene of defeat was perfectly realized by Edison's return to the depot beside the president of the United States. The station itself had been moved to Dearborn to accommodate him.

Not only the station had been moved. The laboratory where Edison and his team created the incandescent electric light and the system of generation, transmission, and installation that made it practical had also been transported to Henry Ford's emerging historical park, located behind his full-scale reproduction of Independence Hall. Yet the laboratory Ford reconstructed was not quite the same as it had been in New Jersey. The restored Menlo Park was presented as an ideal workplace of unalienated labor amid a group of historic buildings transported from many locations.

In 1929 this site had not yet become the famous Greenfield Village tourist site, and many of the buildings there today were not yet in place. What guests saw during Light's Golden Jubilee was its embryonic form.

Attention focused on the authenticity of the reconstructed laboratory. Ford had recovered the abandoned laboratory buildings in New Jersey, dug up their foundations, and transported them to Dearborn, including several railway cars filled with red New Jersey dirt, so the reconstruction could be placed in the same soil. Photographs of the Menlo Park laboratory buildings from 1879 and 1929 show that the buildings were identical in size, form, and arrangement. Edison made the mild objection that the interior was cleaner than it had ever been when in use. The exterior was also more presentable (figure 9.3). The field of weeds had become a trim lawn, and the buildings were freshly painted and surrounded by neat picket fences. The white lampposts of 1929 were an orderly contrast to the rough, unpainted poles of 1879.[37] Aside from these cosmetic changes, the buildings had been recontextualized. Their pastoral appearance merged technology with nature, as if the electrical revolution emerged from rural America. But Edison had only briefly lived in Menlo Park, from 1876 until

Figure 9.3
Thomas Edison's Menlo Park New Jersey laboratory, reconstructed in Michigan at Greenfield Village, 1929. Detroit Publishing Company. Courtesy of the Library of Congress, Prints and Photographs Division.

the early 1880s. It had been a temporary outpost in a rapidly industrializing economy. Then he moved back into New York to construct and run the Pearl Street power station that served lower Manhattan. The large-scale manufacturing required to electrify New York took place elsewhere. Electrification required coal and copper mines and factories producing wires, light bulbs, switches, generators, and motors. The energy transition also required workers to build transmission lines, engineers to run central stations, and a white-collar workforce for new utility companies in every city and town. Menlo Park had produced prototypes that proved the concept was sound. But the growth of the electrical industry and its infrastructure took place elsewhere, and these activities were often disrupted by conflicts between management and labor. Even in the short period of its existence, Menlo Park included a good-sized light-bulb factory that employed more than 150 men, but this facility was not reproduced in Dearborn.

Ford's recreated Menlo Park seemed to have no connection to polluting mines, large factories, or labor conflict. It was animated by nostalgia. The wooden buildings were quaint reminders of another era. The old Edison machine shop on the laboratory's ground floor suggested the artisanal organization of a small workforce where relations were not especially hierarchical, and Edison often labored long into the night with his team. There was an organ in the lab that they played during spontaneous breaks. Separated from its original context, Menlo Park seemed removed from large-scale industrialization. It had been at the cutting edge in 1879, but in Dearborn it appeared archaic. Light's Golden Jubilee omitted any mention of the fact that Edison had lost control of his electrical patents when they were merged into General Electric. Stripped of such facts, what emerged was pastoral and harmonious. The visitors in 1929 could jump over the intervening fifty years of disruptive energy transitions, industrial concentration, monopoly formation, and patent litigation, and see Menlo Park and Edison as the center of a myth, in which industry emerged from pastoral origins, progress was painless, and upward mobility was assured.

These ideas gained credibility from the laboratory's new surroundings. Ford's guests saw not the Menlo Park whistle-stop of 1879 but a village from a generation before then. As the *Detroit Free Press* explained, "the guests passed the old tintype studio on their left. A barker in the door urged them to step in and have their photos made. Next was the post office and apothecary with a post-master and chemist presiding." Several old houses had been moved there as well.[38] Henry Ford had been involved in selecting each building, notably an Illinois courthouse where Abraham Lincoln had once practiced law. It had become a private home before Ford purchased, moved, and renovated it. He had also acquired a school, a country store, an inn, and a boarding house. To complete the village, a work crew built a blacksmith shop, a machine shop, a town hall in the Greek revival style, and a chapel based on the design of a church in Bradford, Massachusetts. For another decade Ford continued to expand Greenfield Village into a display of Americana. By 1940 he had added the Stephen Foster Memorial house from Pittsburgh, the house in Dayton, Ohio, where the Wright Brothers had lived, and the shop where they invented the airplane.[39] The village also acquired the home where the Heinz company originated. These structures were not there in 1929, but it already seemed to be a small antebellum town.

This idealized pre-industrial hamlet contained the seeds of America's future: law, education, religion, work, and invention. While the celebration was called Light's Golden Jubilee, Ford had inserted so many additional elements into the event that it encapsulated nearly a century of idealized American history, stretching the facts of Edison's life to represent the national experience. This history erased any conflict between such villages and industrialization, which was nearby in the gigantic form of Ford's River Rouge plant, the largest factory in the United States. Rather than recognize this contradictory juxtaposition, Light's Golden Jubilee was designed to assert that there was no contradiction between the rural past and the industrial present. Hoover concluded his speech with an extended organic metaphor that also created that impression. "And our civilization is much like a garden. It is to be apprised by the quality of its blooms. In degree, as we fertilize this soil with liberty, as we maintain diligence in cultivation, and guardianship against destructive forces, do we then produce these blossoms, the fragrance of whose lives stimulate renewed endeavor."[40] Edison thus became a flower in the garden of democracy. The ceremonies buried him under such metaphors and associations.

The typology of American heroes had shifted from the frontiersmen like Daniel Boone to the individualistic inventor and the manufacturer. But the economy was still conceived as a natural system that worked best with a minimum of government interference. Just as a beneficent Providence had apparently overseen the manifest destiny of nineteenth-century America, "natural" economic laws seemed to guarantee continued success through invention and manufacturing. Thinking in such terms, selected elements of Edison's life formed an idealized biography. Henry Ford had built a stage where dramatic moments of that life could be reenacted. The illusion that Edison's life had been accurately recreated rested on the authenticity of the setting. The train he worked on as a boy was again operational. Smith Creek Station had been moved to Dearborn. The original Menlo Park laboratory had been restored. The assistant who had helped Edison make the first electric light was again present; and the ceremonial recreation took place after precisely fifty years.

Yet this was a severely edited version of Edison's life. During the half-century since the invention of the incandescent light in 1879 he had pursued other projects.[41] Some succeeded technically but were not commercial successes. His expensive venture in low-grade iron ore mining was not profitable. He developed a system to produce poured concrete houses, but few were made. He struggled to improve automobile batteries, but they remained heavy and took too long to charge, and electric vehicles remained uncompetitive with gasoline cars. His phonographic recordings were technically good, but he had mistakenly assumed its greatest application would be as a tool for business correspondence. Edison lost market share to competitors who specialized in recording popular music. Likewise, despite his patents and early lead in film production, Edison gradually lost his dominant position in that lucrative business as well. In short, his story included economic struggles, mistakes, and losses as well as brilliant successes. But these matters did not shadow the celebration, which reduced his biography to the formulaic story of a poor boy, with little education, who worked hard, invented the electric light, and benefited all mankind.

Simplifying and idealizing his biography made him resemble the heroes described in popular magazines analyzed by Leo Lowenthal. When Edison became famous in the late nineteenth century, journalists celebrated inventors, entrepreneurs, and captains of industry as "heroes of production." They were presented as men who had pulled themselves up by their bootstraps. After 1920 the number of biographical stories in magazines increased, but their focus changed. More than half of them described Hollywood actors, professional athletes, and entertainers, whose rise to fame was less due to hard work than to innate qualities, such as good looks, athletic ability, or an excellent singing voice. They were "heroes of consumption."[42] Edison had emerged as a hero of production, but his patents on the microphone, phonograph, and film projector had helped make media events possible, and in later life he was also a hero of consumption interviewed by reporters for his views on a wide range of topics.

His recordings, lighting, and films had provided Americans with new experiences and forms of consumption that were standardized, packaged, and repeatable. Ford's rebuilt Menlo Park complex was a shrine where these new experiences originated. Edison's restored laboratory became part of a history stripped of its tensions and ambiguities, and his inventions became part of a process in which workers ceased to identify with a craft and instead began to create their identity through consumption and leisure.[43]

Lowenthal sent his study of magazine biographies to Max Horkheimer, who replied, "You will remember those terrible scenes in the movies when some years of a hero's life are pictured in a series of shots which take about one minute or two, just to show how he grew up or old, or how a war started and passed by, and so on. This trimming of an existence into some futile moments which can be characterized schematically symbolizes the dissolution of humanity into elements of administration."[44] At Light's Golden Jubilee Edison's biography was similarly reduced to staged scenes that omitted much of his life.

The day before the Jubilee, when Edison's wife Mina saw the schedule, she sensed something of the sham. Sitting in the Ford mansion where they were staying, Mina wrote a silent protest to the proceedings. "Feel that Menlo Park and Electric light idea is made so prominent that all the other great inventions and work is almost forgotten. General Electric getting more definitely established in having Owen Young toastmaster." She was particularly upset that Menlo Park had become the focal point, believing that the "Orange [New Jersey] laboratories must, and Llewellyn Park must be kept the real shrine." Mina was Edison's second wife, and she first met him after he had left Menlo Park. Everything she had experienced with him for forty-five years was absent from the ceremony. She also had misgivings about the Greenfield Village project: "I do not approve of disturbing historical spots. Let things be." She recognized that the event was "a most magnificent thing of Mr. Ford but I regret the method," for she felt "that dearie is so much more than the electric light and that the light jubilee is one grand advertisement for General Electric and the light companies." She sensed that Edison "has just been made the excuse."[45] Her notes corroborate the disappearance of Edison in the Jubilee despite its scrupulous attention to authentic details. Mina realized that the tribute oversimplified her husband's life. While Ford's admiration was genuine, nevertheless he was helping promote the corporation that had absorbed his most valuable inventions.

Light's Golden Jubilee and Greenfield Village conveyed not the life of Thomas Edison so much as a nostalgic vision, conflating one life with the nation's history. It suppressed reference to larger trends, notably, the alienation of the workers from control of their labor, increasing class stratification, and social tensions, demonstrating how, in Horkheimer's words, "historical information for the masses becomes a cobweb of lies and of ridiculous accumulation of the most insignificant facts and figures" that obliterated complexities.[46]

The development of media events illustrates Habermas's conception of the colonization of the lifeworld by new forms of communication. As he put it, "the subsystems of the economy and state become more and more complex as a consequence of capitalist growth and penetrate ever deeper into the symbolic reproduction of the lifeworld."[47] The radio broadcast of Light's Golden Jubilee was such a symbolic reproduction. It was aired by 140 affiliates of the National Broadcasting Company (NBC). Two well-known emcees introduced the event:

> Good evening, ladies and gentlemen of the radio audience. This is Graham McNamee, with Philips Carlin, speaking from Dearborn, Michigan, where Henry Ford and Edsel Ford are entertaining one of the most notable gatherings assembled in the annals of American history, to honor Thomas Alva Edison on the fiftieth anniversary of his invention of the first practical incandescent lamp. Tonight, is the climax of Light's Golden Jubilee, and what a climax! As you know, the entire world is celebrating this epochal event tonight. Nearly every great city will be aflame, and almost every village and hamlet is having its own celebration to honor the man whose genius made possible this great age of light in which we live.[48]

McNamee urged listeners to "light a candle or sit in the firelight" during the reenactment, until it was time to turn on the lights as a "personal tribute." Citizens were to see their own lives in terms of technological progress. McNamee described the dinner underway in "the exact reproduction of Independence Hall," and suggested that "the world could not celebrate its emancipation from darkness more fittingly than in this building, which is associated in the minds of all of us with freedom and progress."[49] Choosing metaphors that Hoover would reiterate in his speech, McNamee appropriated the symbolic associations that Ford had assembled. Outside the banquet hall in Greenfield Village, Philips Carlin breezily asserted that "life was a pretty proposition in those good old days." He declared that "honestly, I have found out more about our nation's past, what we rode in, what we ate, how we lived and how we amused ourselves, in two hours here today than I think you could get out of a dozen books in a dozen years. You cannot fool me now about American history. I know how we dipped our candles and what we took for whooping cough in the good old days when our grandfathers went courting in snappy runabouts and nobby beaver hats."[50]

When Edison, Ford, and Hoover left the banquet hall for the reenactment in Greenfield Village, Carlin described the incident at Smith Creek Station. Meanwhile, McNamee hurried to the old laboratory. Looking at its walls, lined with bottles containing chemicals of all sorts, he mused, "I spent a lot of time this afternoon reading the labels of these bottles and wondering what they had to do with electricity and lighting. I saw old bottles with their old labels almost rubbed off, reading, "peanut oil," "coconut oil," "turkey red oil," "cedar shavings," "red rock salt," "dogwood bark," and "Jamaica ginger."[51] His list contained only quaint-sounding, natural products, and led up to the reminder that the first long-lasting filament for the light bulb was made from carbonized sewing thread, a familiar household item. The broadcast, like the celebration, focused neither on science, nor technology, nor the process of invention, nor even Edison's life. Rather, it transmitted the sense of participating in history.

Listeners were urged to sit by their radios "under the flickering lights of 1879," until they could rejoice when Edison touched an old battery wire to a reproduction of his original lamp. Then McNamee shouted into the microphone, "It lights! Light's Golden Jubilee has come to a triumphant climax." At this moment the broadcast might have focused on Edison. Instead, to connect the electric light, the American Revolution, and the banquet, the broadcast immediately shifted to the belfry of Ford's full-scale copy of Independence Hall, where a replica of the Liberty Bell was rung for the first time. It tolled out what was described as the victory of light over darkness, while searchlights combed the sky, motorists flashed their lights, planes flew overhead, and Detroit blew all its whistles and sirens. Carlin imagined this tumult spreading to "the rest of the world illuminated by the pealing of the bell," which in his commentary was completely entwined with the light. He intoned, "I can see the beacon of eternal light that has been erected at Menlo Park, New Jersey, flashing on. I can imagine New York, Philadelphia, Chicago, Atlantic City, St. Louis, San Francisco and all the cities of the country, big and small, as they come out of darkness into light. I can imagine all of you about your radios turning on your electric lights and joining in the great celebration in honor of the great hero of peace and progress."[52]

The broadcast asserted that every listener benefited materially (in the use of the light bulb) and spiritually (by participating in the event). By flicking a switch, one became both a consumer of electricity and part of history. As McNamee emphasized, "We are privileged tonight to be present not only at history in the re-making but at history in the making. I hope every child in the country is close to a loudspeaker at this time. Great moments in history are not frequent, and it now seems eminently proper to me that this first international broadcast should convey the thanks of the world to the founder of this age of light." He spoke almost as if he were part of the audience, as if he, too, were just listening in, as if NBC, rather than helping to craft the event, was fortunate just to be covering it.

In the dinner speeches that followed, President Hoover focused less on the light bulb than on "the method of invention."[53] Edison had founded one of the world's first research laboratories, and it was an engine of creativity that made more inventions possible. Hoover extolled research as "the most potent of impulses to progress in our nation." He argued that it ensured a higher standard of living, longer life, more leisure, "more stability of employment," and "more joy in life." He also pointed out that inventions are taxable. The organic metaphors at the close of Hoover's speech asserted that American development was like the harmonious unfolding of a garden, a vision that seemed plausible amid the idealized past of Greenfield Village.

Albert Einstein spoke more pointedly via a radio hookup from Berlin. After congratulating Edison, he said: "Your construction of the bulb has remarkably promoted the development of a big industry of electricity. The great creators of technics, among whom you are one of the most successful, have put mankind into a perfectly new situation to which it has not at all adapted itself. For the production of goods necessary to life, man hardly needs his muscles today. He does not need to be motor or slave." Einstein recognized that electrification had reached a mature stage of development where inventors would lose their centrality. "The big creators of the coming generation will probably be men of organization, and they will have to see to it that the technical powers will no longer serve war, but the economical safety and freedom of mankind."[54] Einstein saw the complexity of Edison's legacy. He alone alluded to the separation of the electrical industry from the inventor, the need for new forms of economic organization, and the danger that technology could be abused. However, this "perfectly new situation"—the new lifeworld—was scarcely mentioned during the rest of the Jubilee. Einstein spoke in German, and few guests or reporters understood him. The fact of his participation alone remained. He was one more historical element in the celebration, which included messages from the Prince of Wales, the president of Germany, Guglielmo Marconi, and Commander Richard E. Byrd at his exploration station in Antarctica. All lent prestige to the celebration, as did famous banquet guests, including Marie Curie, John D. Rockefeller, Will Rogers, George Eastman, Orville Wright, and many executives from the automotive and electrical industries.[55]

Edison's two-minute speech was another example of the absence of meaning. The fact of his presence lent solemnity to his words. But the speech was filled with cliches such as "my deep appreciation," "countless evidences of good-will," "the bottom of my heart," "this unforgettable night," "fills me with gratitude," "words are inadequate to express my feelings," and so forth.[56] Edison did not write the speech. His wife, personal secretary, and son Charles had each prepared a draft. Not surprisingly, given her view of the proceedings, Mina's was the shortest, while the secretary's was the longest and most formal. Edison chose Charles's version and read it almost verbatim.[57] By adopting this speech, he was treating the Jubilee as a media event, forfeiting the spontaneity and humor he displayed in birthday interviews. It proved to be a wise decision, because that evening he felt ill. After reinventing the light bulb, he was so exhausted that he had to rest on a couch before returning to the banquet hall to deliver his brief remarks.[58]

One searches the entire Jubilee of Light without finding a single personal expression from Edison. His words were written by someone else and, like the laboratory, possessed only a specious facticity. The public ritual celebrated more than the invention of a practical incandescent light. Its staging in a replica of Independence Hall conflated Edison's invention with American political values, identifying the "conquest of night" with the Liberty Bell and the spread of democracy. William Barstow, one of the Edison pioneers, declared at the banquet, "In place of empires you have founded industries which are engaged throughout the world in advancing civilization. The struggle of light against darkness has not been won without a signal leader."[59] The event depicted American history as a harmonious unfolding from a village economy to an urban industrial state to world leadership. It presented Edison's laboratory as a symbolic site that guaranteed future transformations.

In the kaleidoscopic imaginary, Edison was the perfect figure at the center of a 1920s media event. He was a white man. He had grown up in a small town. He had little formal education and was largely self-educated. He became internationally famous after inventing the phonograph at the beginning of the energy transition in 1877, and inventions poured forth from his laboratories for the next half-century. Moreover, these inventions were at the center of the growing culture of consumption. Entire industries emerged from or were transformed by his patents, including the phonograph, improvements in both the telegraph and telephone, the practical incandescent light and its system of generation and distribution, the film industry, improvements in the production of glass and cement, X-ray machines, and spectacular lighting displays first seen at world's fairs, as well as improvements in meters, batteries, vacuum tubes, and many other devices. Edison simultaneously represented both individualism and industrialism, both the small town and the metropolis, both the practical tinkerer and science. But Light's Golden Jubilee focused on his electrical patents alone, framed by his small-town origins, self-reliance, and tireless energy.

In 1935, Leonard W. Doob concluded in *Propaganda* that Light's Golden Jubilee was "one of the most lavish pieces of propaganda ever engineered in this country during peacetime."[60] Afterward, the idealized Greenfield Village became a nostalgic monument to an imagined past. As for Edison, when he passed away in 1931 the newspapers lavished praise on him as the ideal, individualistic, white American, born to a modest family in a small town. The other Edison narratives once described in the newspapers—Alger hero, robber baron, entrepreneur, failure, genius, tinkerer, professor, and wizard—had been simplified into an icon.

After Edison's death, there was a proposal to pay him tribute by briefly turning off the lights across the nation. That too would have been a media event. However, President Hoover realized it could have fatal consequences, and he released the following statement:

> The suggestion had been made that the electrical current at generating plants should be turned off . . . [but] this would constitute a great peril to life throughout the country because of the many services dependent upon electrical power in protection from fire, the operation of water supply, sanitation, elevators, operations in hospitals and the vast number of activities which, if halted even for an instant, would result in death somewhere in the country. It is not, therefore, advisable. This demonstration of the dependence of the country upon electrical current for its life and health is in itself a monument to Mr. Edison's genius.[61]

Light's Golden Jubilee had presented Edison as a self-made individualist, but his inventions had been used to intensify media events and to make society more interdependent. He was at the center of the energy transition. His creations had separated light from fire, divorced sound from presence, and made aural and visual experiences repeatable. The nation was not merely tied together by the electrical network, it could not be safely unplugged. Media events made manifest what was rapidly becoming a new normal: A nation of more than one hundred million people was so wired together that everyone from coast to coast could listen to the same news, hear the same songs, and experience events at the same time. As a networked nation emerged, interdependence was replacing individualism.

Three days after the Jubilee, the New York Stock Exchange began to fall and in just two days lost a quarter of its value. The instantaneous communications used at Light's Golden Jubilee struggled to keep up with the feverish transactions, while network radio coverage spread uncertainty and fear to all parts of the country. For the next three years, the market made small rallies but mostly declined, and the Dow Jones Industrial Average in the summer of 1932 was 89 percent below its peak in 1929.[62] The crash signaled the end of an era.

10

THE MACHINE AGE

By 1929 the transition to a high-energy society had passed through the five stages that Hughes identified in *Networks of Power*. That process built upon the growth of steam power and developed the uses of oil, gas, and electricity after 1865. First there was a remarkable series of inventions in the 1870s and early 1880s, including the telephone, improved electric dynamo, wiring systems, and new kinds of electric lights. These discoveries stirred public interest. Crowds came to see the first arc lights and incandescent lights in the 1870s, to see the first automobiles in the 1890s, and to see the first airplanes in the 1900s. In the second stage, the new energy technologies were displayed in public places and demonstrated at expositions in Chicago (1893), Omaha (1898), Buffalo (1901), St. Louis (1904), and San Francisco (1915). The expositions aroused both utopian hopes and dystopian fears, as they showcased spectacular lighting, automobiles, new consumer products, and new factory systems such as the assembly line.[1] In the third stage, development confronted "reverse salients," such how to transmit power over longer distances or how to build oil pipelines. Extravagant expectations began to be replaced by practical issues and by comparisons of competing solutions (Electric or gasoline cars? AC or DC? Mass transit or automobiles?) There was considerable public debate about whether transit companies and utilities ought to be public or private corporations, about whether streets ought to retain a variety of functions or become exclusively transportation arteries, about what regulations and safety standards were necessary for motor vehicles or in factories and mines, about appropriate levels of lighting, and about pollution. As these choices were made, the fourth stage of technological momentum began to spread the new energy systems throughout society, and the new technologies were naturalized. By 1929, the energy transition had entered the mature stage and was becoming the almost unconscious basis of a new lifeworld.

Figure 10.1
Louis Lozowick, *New York*, 1925, lithograph. Courtesy of the President and Fellows of Harvard College.

And yet, in the midst of these transformations, Americans retained some values and narratives from their rural past. They gloried in their founding fathers. They idealized rural life and the small town. They embraced the new national parks. They knew that the "frontier era" had ended, but books that romanticized settlers in the West were still best-sellers. They reconceived frontier individualism to include inventors, entrepreneurs, professional athletes, and others who appeared to be self-made. They still believed in self-reliance, individualism, and positive thinking, and they were convinced that new opportunities existed in the cities. The most widely read author in Muncie's public library at the beginning of the twentieth century was Horatio Alger. His rags-to-riches stories were borrowed more than three times as often as the combined total for the works of Nathaniel Hawthorne, William Dean Howells, Henry James, and Mark Twain.[2] The ideology of self-reliance and individualism was not abandoned but reconfigured. The nation identified with nature and small towns even though it had become urban. It defined itself as a democracy, even though its commercial life was dominated by private corporations. Americans saw their history as a story of unfolding progress, and the rapid transformation of the great energy transition was conceived not as a disruption but rather as a kaleidoscopic continuity. The metaphor proved useful during the shift from a regional to a national economy. It defined change as disconnected images, like a stack of colorful postcards. Kaleidoscopic variety was also celebrated in local color writing about different regions, whether the islands of Maine, the farms of the Midwest, the settlement of the desert West, or the plantation South. In the 1920s these kaleidoscopic evocations persisted, but they were less frequent.

Throughout these years Americans sought out disorienting sensations, available on Ferris wheels, on carousels, in elaborately lighted dance halls, or when viewing the swirling electric fountains, spectacular fireworks, panoramic views from skyscrapers, and other forms of the technological sublime. In the 1920s they also threw themselves into the gyrations of the Charleston and other fast-paced dances, moving to 200 or even 250 beats per minute. The dance emerged from African American dance practices in the rural South near Charleston. It developed further as it became popular in Harlem. The Charleston captured public attention in New York when performed by an all-Black cast in the musical *Runnin' Wild*. The signature tune, written by James P. Johnson, employed a rhythm based on Gullah/Geechee spirituals and the Lowcountry shout or "ring shout."[3] The diaspora of African Americans moving to the North brought this tradition into mainstream culture, where it coalesced with jazz. The Charleston became the signature music of the 1920s. It was incessantly played on phonographs and the radio, and its quick, syncopated rhythms inspired a generation of white and Black youth. It was part of a larger cultural rebellion against straitlaced Victorian norms. In speakeasies and dance halls, it was inseparable from widespread defiance of Prohibition.

During the same years, in hundreds of amusement parks whirling mechanical rides disintegrated experience into a pastiche of energetic fragments.

Decades of using kaleidoscopes in the parlor had to some degree prepared the public for such experiences. Perhaps it also helped them to see the canvases of futurism and cubism. Joseph Stella represented the visual cacophony of mechanical rides in a painting from 1913–1914, *Battle of Lights, Coney Island, Mardi Gras* (figure 10.2). He recalled, "I felt I should paint this subject upon a big wall, but I had to be satisfied with the biggest canvas I could find. . . . I built the most intense arabesque that I could imagine in order to convey in an [*sic*] hectic mood the surging crowd and the revolving machines generating for the first time not anguish and pain but violent pleasures. I used the intact purity of the vermillion to accentuate the carnal frenzy of the new bacchanal and all the acidity of the lemon yellow for the dazzling lights storming all around."[4] As John Kasson explains, *Battle*

Figure 10.2
Frank Stella, *Battle of Lights, Coney Island, Mardi Gras*, 1913–1914. Courtesy of the Yale University Art Gallery.

of Lights presents "a barrage of sensations and emotions." It is a "turbulent scene, a riot of swirling colors and shapes exploding in all directions. Dark-hued crowds surge upward from the bottom of the canvas toward the brilliant carnival," while in the upper part "dancing figures sway sensuously, outlined in pure vermillion." The painting expresses an "exhilarating sense of release, of ecstatic disorientation, of dynamic activity and celebration. Yet there exists a sense as well that the 'carnal frenzy' and 'violent, dangerous pleasures' may escape bounds. The hectic, propulsive rhythms allow no rest. The chaotic activity is almost suffocating in its profusion."[5] The kaleidoscope of urban life turned so swiftly at the amusement park that it overwhelmed visitors with contradictory impressions, mechanized movements, and violent experience.[6]

F. Scott Fitzgerald also found inspiration in the amusement park. In the mid-1920s, in the story "O Russet Witch" he explored the idea that American life had become an accelerated merry-go-round. When writing it, he "was somewhat carried away by the feeling that there was no ordered scheme to which I must conform." Disorder was embedded in the story whose protagonist felt he was on a carousel spinning toward ennui and death. "The years between thirty-five and sixty-five revolve before the passive mind as one unexplained, confusing merry-go-round. True, they are a merry-go-round of ill-gaited and wind-broken horses, painted first in pastel colors, then in dull grays and browns, but perplexing and intolerably dizzy." In this mechanistic vision, adults were trapped in the rotations of modern life, gradually growing confused. This had not been the case on "the merry-go-rounds of childhood or adolescence" or on "the certain-coursed, dynamic roller coasters of youth." But during the thirty years of later adulthood came a gradual withdrawal from life, until "we sit waiting for death."[7] Fitzgerald imagined modern adulthood as imprisonment in a spinning machine, where one became feeble and disoriented, caught in a vortex of unrelenting pressure. This was a vision of entropy, as personal energy declined and disorder increased.

In contrast, most of Fitzgerald's contemporaries still saw themselves as observers of kaleidoscopic multiplicity and change. They lived in a high-energy society, but their imaginary remained that of a universe of force, as described by Herbert Spencer, in which order and complexity increased over time. This was a world of cause and effect with no losses, as force was always being preserved. One was not trapped inside an entropic machine, but an observer of technological progress. Society did not descend into chaos but was constantly improved. Kaleidoscopic vision embraced a succession of views, confident the larger panorama was harmoniously shaped by the invisible hand of the economy. Each new landscape and each news item was part of an ongoing advance, as "the machine age" moved toward inevitable outcomes. One was not traveling in circles, nor tied down to a particular place. Rather, Americans were exploring an emerging perfection, as they improved nature, made industry more efficient, and refined society. Americans were accelerating beyond the local toward a mechanical, national future. In this high-energy society, they grew up in one place, got an education somewhere else, and took jobs in new locations, while friends and relatives also kept moving about. The local became temporary, the citizen almost peripatetic.

To travel through a kaleidoscope of scenery, monuments, and urban landmarks did not call for historical or economic analysis. The parts were related to the whole in the geographical sense that all were in the United States, but in the kaleidoscopic viewfinder there was no hierarchy, no center, no dominant or subservient parts, and no explicit narrative other than the general assumption of progress. Like a random stack of postcards, each new scene was appreciated for itself alone. Kaleidoscopic sites seemed independent of one another, and citizens appreciated a magnificent skyscraper, a mountain view, a subway ride, a battlefield, a small town, or a grove of redwoods as just so many views. There seemed to be no deleterious cause and effect, no exploitation of labor or resources, no class tensions, and no entropic losses. Kaleidoscopic views were not part of a historical agenda but enjoyable episodes. The metaphor of the kaleidoscope suggested an observer looking at a continually transforming scene in a universe of force. Only a minority, like Fitzgerald, adopted the entropic view that Americans were trapped in machinery whose constant movement blurred the surroundings until they became dazed.

In the kaleidoscopic imaginary, Americans were like the young man described by John Dos Passos in the opening pages of his trilogy *U.S.A.* They had walked fast through the streets, eager to take up all the jobs and see all the sights. Between 1876 and 1929 they hurried through the crowds, the cities, and the cavernous train stations. Their minds buzzed with hopes, their muscles stretched toward a cornucopia of jobs, laying down the new asphalt roads and rolling them smooth, fishing from trawlers on Lake

Superior or off the Maine coast, riveting the steel frames of New York skyscrapers, driving the new interurban cars from Muncie to Indianapolis, building bungalows in the orange groves outside Los Angeles, selling electric irons and toasters to Philadelphia housewives, drilling for Texas oil, excavating the Panama Canal, selling radios on Main Street, making a precisionist painting, or writing a modernist novel about Chicago. It was a white man's world of hopeful struggle. They rode the rails, walked the cities, joined the parade, cheered in stadiums, explored new national parks, and searched for their place in the turning American kaleidoscope. It was a continent of possibilities, an unending newsreel, and yet often a journey to disappointment.

Dos Passos developed a kaleidoscopic form for the novel. *U.S.A.* was constructed of many perspectives, including fifty-one subjective "Camera Eyes," short biographies of twenty-seven historical figures, sixty-eight fragmentary "newsreels" of current events, and the personal stories of twelve characters who seldom met one another. The fragmentation of narrative was not unique to Dos Passos, as other writers in the 1920s also developed ways to shatter realism into new literary forms, including Jean Toomer in *Cane*, Sherwood Anderson in *Winesburg, Ohio*, and William Faulkner in *The Sound and the Fury*. In *U.S.A.*, four narrative methods collide and diverge, creating networks of relations between public life, political events, and private experience. As Dos Passos put it, the representative man wanted to "catch the last subway, the streetcar, the bus, run up the gangplanks of all the steamboats, register at all the hotels, work in the cities, answer the wantads, learn the trades, take up the jobs, live in all the boardinghouses. . . . One bed is not enough, one job is not enough, one life is not enough."[8] They restlessly pursued happiness in every cranny of the nation, but they could only partially realize their aspirations. As Fitzgerald suggested of Jay Gatsby, some foul dust floated in the wake of their dreams. They could not all succeed. They could not make a more perfect union when millions of farmers struggled against low prices in the unelectrified countryside beyond the zest of the cities. Nor was there liberty and justice for all in a nation that excluded women from many professions, segregated Black Americans, stranded Native Americans on reservations, and excluded Chinese and Japanese immigrants. Even for white men, kaleidoscopic America pandered to a desire for success that many could never attain.

At the end of *The Big Money*, the third volume of *U.S.A.*, Dos Passos's young man was worn out and alone, with no job, no family, and no home.

He was hitchhiking west, wearing a threadbare suit and broken shoes. When young, the world had seemed wide open, and he had traveled far and wide to take it in. During the great energy transition, the opportunities had seemed limitless. But during the Great Depression the blinders came off. Individualism was not sufficient. Progress was neither assured nor automatic. The nation was not a series of postcard views, but a tightly interwoven network. He stood alone on the roadside. He was hungry and "wants crawl over his skin like ants."[9] He belonged nowhere, and none of the passing cars offered him a ride.

In the *U.S.A.* trilogy the kaleidoscopic form was deployed to criticize the intensification of industrial civilization. As Lisa Nanney emphasizes, Dos Passos struggled to comprehend the meaning of mechanization in American life and to find an adequate artistic form even before he graduated from Harvard. In "his final essay for the *Harvard Monthly*, he indicted industrialism not only for not affording humankind with greater creative freedom but for actually '[binding] three-fourths of the world . . . in economic slavery' as laborers producing goods so that the 'other fourth may be enslaved' by consumer capitalism." In the 1920s he wrote plays that "dramatized his concern with the dehumanizing effects of industrial capitalism and the creation of a hollow myth of success in an America where economic inequality increasingly widened."[10] Yet, like Stella, he sensed a creative potential in modern machinery and was attracted to Futurism.

Modernist painters fragmented their works to suggest the flux and multiple possibilities of the new lifeworld. This artistic interest culminated at the 1927 Machine Age Exhibition in New York. Writing for its catalogue, Louis Lozowick saw artistic possibilities in "the rigid geometry of the American city, in the verticals of smokestacks, in the parallels of its car tracks, the squares of its streets, the cubes of its factories, the parts of its bridges, the cylinders of its gas tanks."[11] American painters, sculptors, and photographers, he argued, were transmuting these structures and rhythms into art. The works of Morton L. Schamberg, Francis Picabia, Man Ray, and Marcel Duchamp often focused on machines.[12] Inspired in part by cubism, they represented machines in collage-like patterns or as disassembled parts. Schamberg produced abstract works based on machines (figure 10.3); Charles Sheeler presented factories as formal compositions (figure 10.4); and Marcel Duchamp played with the forms of mass production. Other artists explored factories, airplanes, skyscrapers, and spectacular lighting

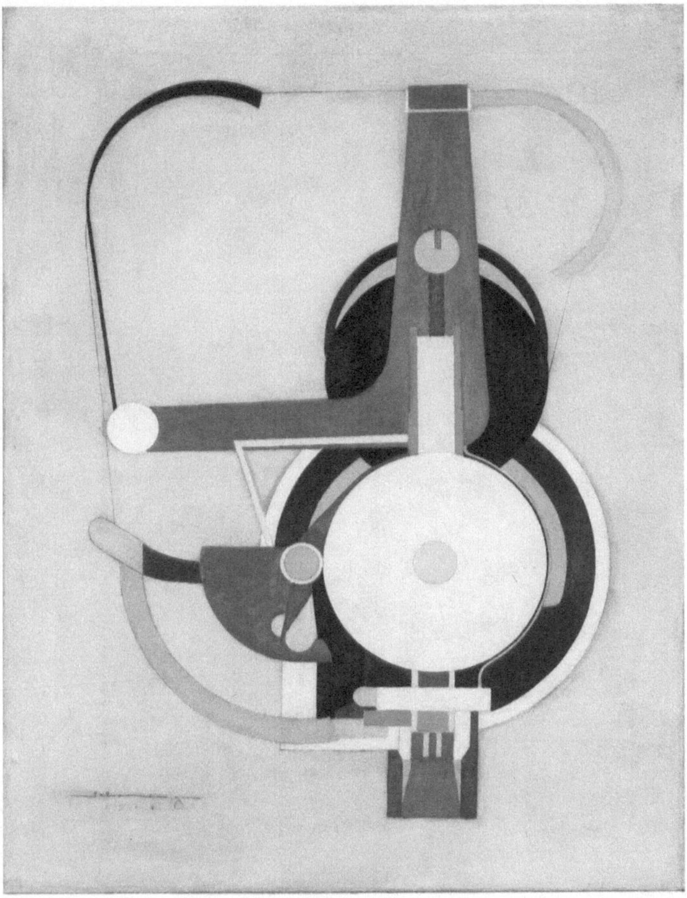

Figure 10.3
Morton Schamberg, *Painting* (formerly *The Machine*), 1916. Courtesy of the Yale University Art Gallery.

effects. The machine aesthetic was also appropriated by commercial artists and designers, and by the mid-1920s it was evident in advertisements for Chrysler cars and other consumer goods,[13] as well as in skyscraper design and ornamentation.[14]

Artistic interest in the machine had intensified as the great energy transition entered its mature phase. A majority of factory power had shifted from waterwheels and steam engines to fossil fuels or electric motors. The new energies also were spreading rapidly to every household, not only in the cities, suburbs, and towns but in the more prosperous rural areas. As this

Figure 10.4
Charles R. Sheeler, *Industrial Architecture*, 1931. Conté crayon and graphite on paper mounted to board. Courtesy of the Amon Carter Museum of American Art, Fort Worth, Texas.

process accelerated, the kaleidoscopic metaphor declined in use, and Americans increasingly explained their changing society by referring to "the machine age." This phrase did not refer to a specific machine or mechanical process but rather to mechanization in general, including oil wells, refineries, pipelines, power plants, factories, or aesthetically interesting machines. The phrase "machine age" was common; the word "technology" was not.[15]

Americans did not perceive the great energy transition but rather its myriad effects. In part this was because they had seen continual battles between gas and electric companies in street lighting, between coal and oil in the heating market, between steam and electricity in industrial markets, or between mass transit and automobiles. There was not one energy company but many corporations in competition. Furthermore, there seemed to be a great difference between dirty coal and clean electricity, which consumers connected with dams, even though it was more often produced by coal-fired power plants.[16] In contrast to steel mills and factories that spewed black smoke, electricity could be incorporated into a pastoral vision of harmony with nature.

Americans lacked adequate terms to think with. As Schatzberg explains, "By the end of the nineteenth century, the terms art and science had lost most of their utility for explaining changes in what we now call technology. This loss occurred in part because scientists and fine artists" had narrowed their self-definition to achieve "professional autonomy. Art was aestheticized, while mechanical and industrial arts were debased. Science was made the property of a small elite." Consequently, Americans failed to understand "the Second Industrial Revolution, the period from about 1870 to 1930, perhaps the most profound technological transformation in human history." Worse, "the net result of these conceptual changes was to remove human agency from the discourse of industrial modernity."[17] Americans tried to fill this semantic void, that "technology" would fill, with "the machine age." The word "technology" when it emerged, borrowed some of its meaning from the German *Technik*, notably as developed by Werner Sombart. He rejected a technological determinism and recognized that *Kultur* shaped material change.[18] He and other German authors influenced the young Lewis Mumford, who spent months in 1932 reading in the library at Munich's Deutsches Museum. His seminal *Technics and Civilization* (1934) rejected deterministic conceptions of technology and argued for the importance of culture and human agency.[19] But this non-deterministic understanding of technology was absent in the popular 1920s concept of "the machine age."

Stuart Chase was one of the few who sensed the semantic void that made it difficult to name, much less analyze, the great energy transition. In *Men and Machines*, he commented, "It has been called the Iron Age, the Age of Steam, the Coal Age, the Machine Age. Lately we hear, depending on the trade association involved, the Oil Age, the Electrical Age and the Motor Age. In a year or two, the Air Age will be a well-worn phrase." Rejecting all of these, Chase declared it a "Power Age" that began with Watt's improved steam engine 150 years before.[20] This conflation of the first and the second industrial revolutions was the best anyone in the late 1920s seemed able to do.

Chase also was one of the few to adopt the word "technology," although he had conflicting definitions of the term. He identified it with mechanical devices and devoted an entire chapter to analyzing the effects of the seventy most common machines of the 1920s. But he lacked a comprehensive theory.[21]

The idea that history might consist of a succession of technological lifeworlds came closer to being realized in Mumford's *Technics and Civilization*. He disputed the idea of linear progress, for example, arguing that industrial life in the later nineteenth century was a time when "the eye, the ear, the touch, [were] starved and battered by the external environment," and therefore people learned to experience the world "second hand" through guidebooks, museums, and other representations. He concluded that in Victorian England "starvation and diminution of life was universal," fostering "a state of partial anesthesia" that "became a condition of survival."[22] This anesthesia was not inevitable, for Mumford saw not a deterministic machine age, but individual machines, each of which was "an instrument with manifold and conflicting possibilities." A society could make unnecessary and shoddy products, and it could "counterfeit older forms of art," but it also could use new machines "to concentrate and intensify and express new forms of experience."[23] Like Dos Passos, Mumford recognized that a new machine aesthetic, developed by engineers and artists, had come to formal expression. It valued "calculation, flawlessness, simplicity, and economy."[24]

However, such nuanced analysis was absent during the 1920s. Instead, most Americans adopted an optimistic ideology of progress and believed in a deterministic narrative of second creation, in which Americans used new machines to continually improve society.[25] Progressive political movements assumed this narrative of automatic advance, in which machines raised manufacturing efficiency, eliminated child labor, improved sanitation, mitigated smoke pollution, regulated traffic, industrialized agriculture, and linked every household in networks of services. This was the dominant narrative of American progress, and the rapid economic growth of the 1920s seemed to confirm it. In this narrative, regional differences declined, society became more secular, and the population became more homogeneous as they adopted mass production and mass consumption.

Determinism also underlay critiques of "the machine age." In 1927, *The Living Age* complained, "Ever since the growth of the machine age, there has been a destruction of natural beauty, a reckless squandering of irreplaceable natural resources, and a progressive loss of all the exquisite works of skill."[26] When Charles Beard put together a series of essays by well-known contemporaries, under the title *Whither Mankind: A Panorama of Modern Civilization*, the authors mentioned "the machine age" forty-seven times, while analyzing science, business, labor, government, the arts, philosophy, play, education, and literature. They did not all mean the same thing by "the machine age," but they recognized that a comprehensive change had occurred. As John Jordan later concluded in *Machine-Age Ideology*, for Beard, more than some of his contributors, "the machine—infallibly logical and superseding mere humanism—remained the source of hopefulness."[27]

Yet "the machine" was not a carefully defined concept. It was more a gesture than an explanation. For example, Floyd Dell argued in *Current History* that "the old-fashioned home was shattered by the machine age" and that "feminism is essentially an effort of women to adjust themselves to the new conditions created by the advent of machinery."[28] How precisely "the machine age" had transformed the home was not elucidated, nor did he explain why he thought feminism was essentially an adaptation to mechanization. For Dell, like most others, "the machine" became the all-purpose agent of change. Critics expressed anxiety about "the machine age" in government reports and the halls of Congress. The Department of Labor declared in an analysis of technological unemployment that "this age can properly be called the machine age," and went on to analyze the plight of those displaced, "who must be absorbed into other lines."[29] In 1928 a congressman from New York City declared in the House of Representatives, "The twentieth century of civilization will be known as the great machine age in which most of the industrial work upon farms and elsewhere is being perfected and done through the medium of machinery." This had many benefits, but Representative William Sirovich was concerned about older people made redundant by new machinery. They had such difficulty finding work that some committed suicide. He argued that because mechanization rendered so many unemployed, a national pension system was necessary.[30]

Edward A. Filene took a more hopeful view in *The Way Out*. He recognized that it had become "the fashion in theoretical circles to cry out against machines and against all the industrial policies they make necessary." It appeared that "our machine civilization hasn't a chance before the bar of current theory. It is adjudged guilty on every count."[31] Filene tried to convince readers that working hours would decline to about five hours a day, that factory jobs could be made less monotonous, that consumer demands for variety and quality would force corporations to improve products, and that the industry of the future would be decentralized into smaller, dispersed factories in the countryside.[32] But such unverifiable predictions did not silence a critique that Richard Popp nicely summarizes: "Radios, cars, films, and other innovations seemed to draw distant places closer, while at the same time scattering listeners, motorists, and viewers further afield."[33] These machines "entered everyday life under the guise of play, but once they were made routine, they monopolized private hours, overwhelmed community-sustaining forms of leisure, filled citizens' minds with inane concerns, and conditioned popular tastes toward easy gratification over authentic experience; rather than offering means for human autonomy, they instead provided avenues for commercial expansion." The result was absorption into "a phantasmagoric national culture that was at once everywhere and nowhere."[34]

American mechanization both fascinated and appalled foreign observers. In France, André Siegfried wrote a best-selling book that recognized the tremendous productivity of the United States, which provided material welfare superior to that in Europe. Yet he bemoaned the "tragic price" exacted by standardization. In his view, Americans were becoming automatons. The worker was but "a cog in a vast machine." Siegfried was somewhat optimistic compared with the authors of a spate of other French books that attacked "the American menace."[35] In Britain, as early as 1901 W. T. Stead's best-selling *The Americanization of the World; or, The Trend of the Twentieth Century* focused on the rising standard of living and the adoption of mass-produced products.[36] Two decades later, the British novelist Aldous Huxley reflected on "the machine age" in *Harper's*: "The future of America is the future of the world. Material circumstances are driving all nations along the path in which America is going. Living in the contemporary environment, which is everywhere becoming more and more American, men feel a psychological compulsion to go the American way. Fate acts within and without; there is no resisting. For good or for evil, it seems that the world must be Americanized." That meant mechanization in all aspects of life. Huxley continued, "Machinery, then, has created leisure and multiplied the number of impressions which men and women can receive. Universal leisure and variety of impressions make possible a rich universal culture. Machinery has set up a tendency towards the realization of a fuller life."[37] Yet Huxley realized that human beings might prefer enjoyment more than mental cultivation. The mechanization of leisure might mean fewer people could draw a picture, play a musical instrument, spin a yarn, or entertain themselves. The contours of this new lifeworld were not yet clear.

While Americans did not conceive of an energy transition, they knew they were living through an era of rapid change. One of the most widely read works of social science from the 1920s was Robert Lynd and Helen Merrell Lynd's *Middletown*, a sociological study based on years of field research in Muncie, Indiana. It was chosen as a representative community from among 143 cities whose mix of ethnic groups, level of industry, and growth rate mirrored national averages. However, Muncie was atypical, in that it had few immigrants. This was not an oversight but part of the research design. It was decided that including recent arrivals would complicate the study of the "average" American. Muncie did have a small Black population, but they too were intentionally ignored in the research, and the resulting book described the city as though there were no immigrants or African Americans. Middletown epitomized the kaleidoscopic ideal of a city that was entirely white. In the 1880s it was an agricultural county seat of 6,000 residents that suddenly industrialized after natural gas was discovered nearby. By the 1890s its population had doubled, and by 1920 it was 35,000.[38] This was similar to the growth rate in the large cities discussed in chapter 2. Middletown's three large factories each employed between 1,000 and 2,000, and eight other plants had payrolls of 300 to 1,000. These industries produced automobile parts, glass jars, silver-plated tableware, and small appliances. Most of these companies had not existed in 1880.[39] They automated manual labor such as blowing glass jars, and they adopted assembly lines for much of the production.[40] Work had been reorganized by using inexpensive fossil fuels and electrical motors, moving belts, heat sensors, shut-off devices, and other forms of automatic control. When the field work for *Middletown* was completed in 1925, the city had reached the mature stage of the energy transition. Almost all of its 9,200 homes were electrified; the city's Union Traction carried sixteen million streetcar passengers a year; the town had a popular radio station;[41] and its residents owned around 6,500 automobiles, while the surrounding county had another 5,500.[42] Middletown typified the great energy transition.

However, *Middletown*'s authors ignored oil, coal, and steam power completely, and made only passing reference to natural gas or electricity. This blindness was built into the research design, based on functional anthropology. The book's chapters were organized into six sections: work, domestic life, education, leisure, religion, and community activities. The Lynds, like other commentators in the 1920s, discussed not energy systems but individual machines, notably, the automobile, radio, telephone, motion pictures, and electrical appliances. They recognized that using these new devices overturned traditional customs of courtship, modes of thinking, entertainment, and domestic habits. They came close to recognizing that Muncie's lifeworld in 1925 was radically different from what it had been in their baseline year of 1890. Had they been thinking in terms of energy transitions, their baseline would have been 1875, just before the invention of the phonograph, telephone, and incandescent electric light. Much of the book's continuing interest arises from contrasts between the lifeworld of late nineteenth century and that of the 1920s. In 1875, communication was face to face; few people had carriages and most walked around town; long journeys were by train; the horizon of the news was local, as almost no one read a newspaper printed elsewhere; and radio did not exist. Newspaper advertising in 1880 had more text than images. It became more visually arresting by 1925, and it covered six times as much space. There was more advertising because Muncie's residents had higher real incomes, making it possible to buy automobiles and appliances. No family had an electric iron in 1880; 92 percent did in 1924. They had more leisure time, but they no longer wanted to hear lectures of one or two hours, as had been popular in 1890, nor did they attend union meetings as often.[43] Once-popular choirs and singing societies had all but disappeared in 1925. Instead, people went to the movies, listened to the radio, or took a drive. In every sphere of life, habits and expectations had changed. In the new normal of 1925, the national culture was becoming dominant. In 1876 the people of Muncie could not hear a voice unless the speaker was present, but in 1925 distant voices reached them from the telephone, phonograph, and radio, and events from the past and from far away were visible in films. Active participation and face-to-face encounters were in retreat, and passive participation was becoming more common. In the new lifeworld one was less a performer than part of the audience,[44] and the entertainment often was recorded elsewhere. The overall effects widened horizons but diminished local engagement.

The middle-class family depicted in Muncie's advertising was enmeshed in networks of high-energy technologies. Their level of comfort improved as they acquired central heating, hot and cold water, indoor flush toilets, and connection to a sewage system. New electric machines transformed their domestic space, including vacuum cleaners, water heaters, toasters, thermostats, kettles, and other appliances. Electric light was safer and brighter than oil lamps or gas mantles. Electric irons were cleaner and easier to use than old sadirons that had to be heated on a stove, making their handles hot and quite possibly picking up a little soot. Washing machines ended back-breaking scrubbing. A family in 1876 had relied on horses; heated only parts of the house with stoves and fireplaces; fetched water from outside; relied on candles, lanterns, or gas for lighting; relieved themselves in a privy; and lacked immediate access to news or entertainment. The conveniences and services built into a Muncie home of 1929, when *Middletown* was published, demanded an energy intensity more than double that of 1876. An automobile's horsepower far outstripped the team of horses owned by the family's grandparents.

The new energy sources demanded less time and effort than the old, muscle-powered regime. Horses needed daily attention even when they remained in the stable; a bicycle or an automobile did not. Water hauled up from a well or pumped from an outdoor spigot had to be carried into the house, but indoor faucets could be turned on and off in a second. Electric lights also came on at the flick of a switch, in contrast to either gas or lanterns, which demanded care to light and to extinguish. Burning gas, kerosene, or oil also dirtied glass shades and walls, polluted the air, and posed fire risks. But if new appliances offered convenience, they were used to redefine gender roles and transfer much of the housework from the rest of the family to the mother.[45]

The energy transition was not evenly distributed. Busy, brightly illuminated cities like Muncie contrasted with the unelectrified hinterland. Most of the best jobs were reserved for white men and were not available to women or Black people. African Americans often had muscle-powered jobs such as shoveling coal in boiler rooms, washing and ironing clothing, digging ditches, or cleaning offices. They were subjected to segregation and the threat of violence, for Muncie had a chapter of the Ku Klux Klan, which was quite active in Indiana in the 1920s.[46] Some African Americans with light complexions "passed" into white society; others joined the NAACP to contest discrimination through the courts. Native Americans, Hispanics, Jews, and immigrants from China and Japan were omitted from *Middletown*, in the same years that supposedly scientific studies proved their inferiority. White people also were subdivided, and some were ranked as less desirable. The highest status was accorded to Muncie's "old stock" Protestants and the business class. Likewise, national advertising depicted well-to-do, heterosexual, white, middle-class families as representative citizens whose high-energy consumption established the norm.

Like the sociologists who examined Muncie, Americans did not conceptualize the transformations they had lived through as an energy transition. Most people still did not think in terms of thermodynamics but rather in terms of Newtonian mechanics. They did not comprehend an event like Light's Golden Jubilee in terms of changing energy regimes leading to a new social construction of reality. In 1876 Americans had not imagined that in fifty years they would fly, that they would receive radio signals that carried music and information hundreds of miles, or that the average family would own a "horseless carriage" that could drive forty or fifty miles an hour. They did not imagine that they were to become a society where sights and sounds would be recorded, played back, bought, sold, and mixed. They were not accustomed to electrical metaphors to describe body and mind. They were not "live wires," and their breakdowns were not "mental short circuits." The metaphors of 1876 came from horses or steam engines. People might be thoroughbreds, racehorses, or nags. They might be skittish or have a burr under their saddle. They might feel their oats, and they did not think of their coffee as battery acid. In 1876, an angry person might get up a head of steam or explode like a steam boiler, but they would not get their wires crossed or blow a fuse. New machines provided new metaphors to describe people ("running on empty," "asleep at the switch," "a dim bulb," "electrifying performance") and relationships, which could be "high maintenance," "in neutral," or even "in reverse." The metaphors changed in tandem with the emergence of another lifeworld. The energy transition brought with it reimagination of the self.

Changes in consciousness or identity had not been part of the narrative at Light's Golden Jubilee. It offered a straightforward story of progress achieved through individualism. Its central figures—Edison, Hoover, and Ford—represented inventors, engineers, and entrepreneurs who developed new technologies that benefited society. The idealized small town of Greenfield Village contained the seeds of the industrial present. In the narrative of that ceremony, adopting new technologies did not change the culture or the individual. Listening to it on the radio, it seemed that no jobs were lost, no skills became obsolescent, no group was marginalized, and no landscapes were despoiled by mining for copper, coal, iron, lead, or other raw materials. Greenfield Village stood for small towns that were the cradle of progress, exemplified by the garage where Ford built his first car and the laboratory where Edison invented the electric light. In the 1930s these iconic buildings were joined by the Wright brothers' bicycle shop where they invented the airplane. In effect, both Greenfield Village and Light's Golden Jubilee declared that growth occurred without affecting values or identity. They said that the nation had evolved in harmony with the past. It had been not a transition but triumphant continuity.

Yet in fact American society had undergone a sweeping transition. Its economy had become the largest in world, with a higher standard of living than any other nation. Based on the extensive use of oil, gas, and electricity, it had become a networked society woven together in an interconnected system that included the telegraph, telephone, water and sewage systems, streetlighting, traffic lights, fire alarms, burglar alarms, oil and gas pipelines, the stock market, wire services, and control systems in factories, railroads, subways, streetcars, office buildings, and the modern home. Energy had been used to transform cities, industries, and farms. Cities had installed underground railways, electric streetcars, paved roads, spectacular lighting, department stores, museums, office buildings, skyscrapers, movie theaters, amusement parks, sports stadiums, and rapid communications systems. Factories had abandoned the multistory architecture necessitated by water and steam power and relocated in sprawling one-story, electrified buildings. Assembly line production redefined many jobs, and a worker in 1929 produced more than twice as much as a worker in 1876. Streetcars, automobiles, trucks, buses, and tractors displaced oxen, mules, and horses. Agricultural machines replaced muscle power, which meant larger farms, increasing specialization, and a decline in rural population.

By 1929 the Dow Jones Industrial Average was mostly based on companies founded during the new energy regime. A nation of small businesses had been transformed into one dominated by large corporations. Presidents Roosevelt and Taft broke up some of the most powerful monopolies, but oligopolies still dominated many economic sectors. A few oil companies controlled the oil industry. Ford and General Motors produced most of the automobiles. Westinghouse and General Electric made most of the electrical machinery. Yet these powerful corporations presented themselves to the public as civic-minded public service institutions. Americans had the world's highest per capita energy use, and a good citizen was expected to be an avid consumer. An advertising and public relations industry that had scarcely existed in 1876 promoted conspicuous consumption, driven less by need than by desire, and epitomized by the annual rollout of new model cars.

The high-energy system had woven regional cultures into its networks. As the intensification of energy use linked the country more tightly together, the kaleidoscopic metaphor became less convincing or useful. The United States no longer resembled a series of disconnected picturesque views but had become an interwoven and mutually dependent system. Mass production and "the machine" apparently led inexorably toward homogenization. National celebrities eclipsed local heroes; media events marginalized local affairs; national brands took market share from local products. Hollywood films drew the audiences who previously frequented local theater productions. The word "kaleidoscopic" was still used, but it declined in the popular press.

In contrast, the expression "big business" had scarcely existed in 1876, but by 1905 it was more common than "kaleidoscopic." By 1919 it was used eight times more often.[47] "Kaleidoscopic" suggested a glittering panorama of appealing changes but did not lend itself to analysis. To call something "kaleidoscopic" was to declare it exciting but so complex that it defied analysis. In contrast, "big business" pointed to specific actors and had explicit connotations. In the progressive narrative, "big business" was praised for its efficiency and profitability, but not if monopolies formed, because they suppressed individualism, exploited labor, and created inequality. This narrative praised the scale and efficiency of "big business" but also warned that monopolies thwarted free markets and stifled personal initiative. Conservationists like John Muir charged "big business" with rapacious use of natural resources. Mining companies, railroads, and heavy industry drove species to the brink of extinction, left deep scars on the land, and pumped smoke and gases into the atmosphere. In addition to these environmental costs, big business realigned power relations. Smaller companies were bankrupted, absorbed, or reduced to parts suppliers. Farmers using horses in hilly New England or Appalachia lost out to mechanized farms in the flat Midwest using tractors and other gasoline-driven equipment. The countryside lost political power to the cities. Towns reliant on waterpower could not compete with factories that relied on steam, and steam-driven factories were challenged by more flexible plants based on electricity. In short, "kaleidoscopic" and "big business" suggested different narratives. One celebrated regional diversity, the other, like "the machine age," pointed toward an emerging national homogeneity. The kaleidoscopic metaphor suggested cost-free progress rooted in the concept of force. The associated idea of the conservation of forces further suggested that when consuming energy nothing was lost. In contrast, "big business" and "the machine age" frankly recognized the cost of hydroelectric dams, high-tension lines, irrigation canals, oil and gas wells, pipelines, copper mines, coal mines, iron mines, steel mills, airports, and other new infrastructure. Big business in "the machine age" treated nature as raw material that was being transformed into a second creation.[48]

Whatever the metaphor adopted, the change had been astounding. In just fifty years, new energy sources had increased agricultural power by 800 percent and industrial power by 600 percent. Almost every job was quite different from what it had been in 1876, whether plowing, mining, forging steel, making cement, constructing an office building, assembling a product, or running a company. Every worker needed new skills to operate new machines. The expanded scale of industry also meant that workers lost influence on the shop floor. The need for photographers, architects, dentists, machinists, plumbers, electricians, and many other skilled workers skyrocketed, but demand for glassblowers, coopers, cigar makers, blacksmiths, tailors, carriage builders, or saddle makers radically declined. Many new semi-skilled jobs emerged, as the high-energy economy consolidated the nation. In the tightly integrated market, big business dominated through greater specialization, economies of scale, and intensified communications. Without electric lighting, pumps, motors, elevators, fans, and signaling systems, neither a skyscraper nor a subway system nor a modern factory was feasible. Without instantaneous electrified communications the stock markets and newspapers would remain more local than national. The modern city could not exist without electricity to pump water, to control heating systems and alarms, to regulate traffic, to refrigerate food, to power communications, and to run machinery of all kinds. It required networked systems, artificial light, rapid transport, and instantaneous communication.

Living within these networks, Americans no longer discerned the time and season by observing the sun, crops, and foliage. In 1875, time was local, and it varied slightly from one community to the next. By 1929 not only was time standardized down to a fraction of a second, but the economy had been synchronized. Americans relied on precise clocks and thought in terms of a mechanical absolute. As precise timekeeping permeated tasks, meetings, and appointments, once approximate itineraries became exacting and rigid. The national stock and commodity exchanges opened and closed everywhere at precisely the same moment. The electrical utilities were carefully synchronized as well, so that they could exchange alternating current and balance the load. Timers regulated industrial machinery; thermostats controlled the temperature in greenhouses, bakery ovens, and steel mills; news was broadcast precisely on the hour. Everywhere, the individual was enmeshed in systems that ran on schedules. Standardized time was also expressed in synchronized swimming, the choreographed movements of the chorus line, and the carefully timed tasks on the assembly line. The synchronized time system also made possible national media events, linking millions of Americans to planned, simultaneous experiences. The high-energy lifeworld standardized and micromanaged time. Railroads, automobiles, and airplanes sped ever faster across the increasingly engineered landscape. Life seemed to accelerate at breakneck speed.

Technological systems had become incessant. In 1876 the stove was not always lighted, the horse was not always saddled, hot water was not available by turning a tap, and light was not immediately available at the flick of a switch. The telephone did not exist, so to get information one had to get a newspaper or look for a person who could tell you. To get light, heat, power, or information required preparation, time, and muscular effort. This seemed natural, for it had been like that for centuries. In 1929, energy was available almost instantly; it was not intermittent but constant. The city's intense artificial light made it difficult to see the stars, and city people seldom heard birds or insects because urban machinery hummed and rumbled in a constant background roar, punctuated by sirens and horns. The telephone, the electrical system, the heating system, the fire alarm, the newspaper wire services, and the hot-and-cold water supply functioned around the clock. These perpetually available systems were fast becoming naturalized, as were automobiles, elevators, escalators, airplanes, thermostats, radios, films, and other devices that used electricity, oil, or gas. The high-energy lifeworld of 1929 provided almost instant access to its services, made fewer muscular demands, and enabled Americans to be more efficient and do more every day. The drive for efficiency meant compressing more experience into less time, whether at work, on holiday, at an amusement park, or at home.

The horizon of experience had changed. In 1875, the rest of the world was known primarily through printed media: newspapers, magazines, and books. One knew about the rest of the United States, but its news only intermittently intruded into everyday life. Few people had ever seen or heard any of the presidents, for example. The horizon in 1929 was far more expansive, as radio carried sounds and voices across the continent. A farm family in the Midwest might hear an orchestra playing in New York, a baseball game in Chicago, or a politician in Washington. They could also see newsreels and films. The wider horizon was stimulating. It provided a richer and more varied framework for experience, but it could suggest that a farm, a small town, or even a regional hub, was parochial.

The experience of space also expanded after 1875. For the first time in human history, one could ascend to the top of a skyscraper in an elevator, rise effortlessly on escalators in a department store, hurtle underground in a subway, fly above the clouds in an airplane, or control an automobile at night, riding behind a cone of light that carved out a zone of visibility. New spatial experiences were available at amusement parks, including impressive illusions, such as the destruction of Pompeii or a hurricane destroying Galveston. The new spaces were accessible with little or no effort, for motors, loudspeakers, and other machines did the work. The extensive use of lighting also remade landscapes, which took on distinctive appearances during the day, at twilight, and at night. One impressive and disorienting experience was the view of downtown at night, particularly a site like Times Square. There, as Sandy Isenstadt explains, electricity "translated linguistic particles into pulses of light, rich with signification but without a grammar to organize them." It was a pulsating mass of "indecipherable signification."[49] The lights of Broadway lured the eye to read them up and down and from left and from right, to no avail. Perspective collapsed, for there was no common scale to the signs, and no clear background and foreground. Three-dimensionality disappeared, and the scene changed radically depending on where one stood. An amorphous surface was all that remained, as corporations competed to attract the consumer's eye. A new attention economy had emerged in a lifeworld that could overwhelm perception.

The great energy transition continued after 1929, for Americans were determined to retain their high-energy lifeworld despite the Great Depression. In 1940 they owned 4.4 million more cars than in 1930, and gasoline consumption had increased more than 50 percent.[50] Homeowners bought more appliances, and their demand for electricity increased 80 percent.[51] Rural electrification programs hooked up millions of farms. The industrial and commercial use of electricity doubled.[52] In every remaining decade of the twentieth century, Americans would consume more energy and produce more pollution. The average house kept expanding, and it contained more appliances. In the 1920s a family seldom had more than one car, but by the 1960s it was common for every adult in the household to have one. The pattern for these increases had been established by 1929. There were more single-family homes rather than more apartments, more private automobiles, less mass transit, fewer people in rural areas, more in the suburbs, more mechanization of production to increase efficiency, more white-collar jobs, and more tourism. But ever-higher energy use did not resolve social problems: Class differences persisted, racial and gender inequality continued, unsustainable resource extraction grew, and air and water pollution worsened. By the twentieth century's end Americans had become aware of climate change, but they assumed that such environmental problems could be resolved by a technological fix or two. Americans still did not understand that entropy was inescapable, and they still measured national well-being by the pace of economic growth. In 1925 the high-energy lifeworld seemed the enviable culmination of centuries of progress, and in 2025 it retained enormous technological momentum.

EPILOGUE

Energy transitions do not lead to definite results. They open possibilities. Between 1876 and 1929, the United States underwent the transition to electricity and fossil fuels without even having a name for it. The great energy transition made possible enormous increases in productivity, shorter working hours, higher real wages, longer life spans, faster transportation, and many conveniences. The new energy forms were woven into a culture that needed to heal sectional differences after the Civil War, absorb an enormous influx of immigrants, shift from a rural to an urban society, and adopt a scientific understanding of nature. It was a complex, interactive process. Neither oil nor gas nor electricity caused the triumph of individualistic automobiles over mass transit, nor did they automatically lead to a preference for single-family homes, immigration restriction, high tariffs, or racial discrimination. Yet, by the 1920s these choices might seem to be partially responsible for prosperity.

A century later, the United States is again passing through an energy transition, adopting the new technologies of wind, solar, tidal, and thermal power, in part because they are less expensive and in part because of rising global temperatures. This transition demands new technologies of energy production but few changes in consumption. The electricity coming from a wall socket is the same, whether from a windmill, a coal-fired power plant, or a hydroelectric dam. Electric automobiles are not so different from gasoline cars that they require a new road system or driver training. In the earlier transition, outdoor privies were eliminated, streets were paved, internal combustion engines replaced horses, and electricity became ubiquitous. The current transition demands less obvious changes for consumers. Like all energy transitions, this one took half a century, and it began slowly. From the 1980s until 2010, the new technologies cost more than fossil fuels. But by 2015 they were competitive, and before 2020 wind and solar power were cheaper than coal-fired plants. Yet half the nation resisted

Figure 11.1
Joseph Pennell, *Towers of Oil*, 1917, lithograph. Courtesy of the Library of Congress, Prints and Photographs Division.

the change. In 1926 the United States had been a world leader in the adoption of fossil fuels and electrification. But in 2026 it was divided between those city and state governments that had adopted alternative energies and a laggard federal government that championed the old energy regime. The Trump administration pursued policies much like those of presidents from William McKinley to Herbert Hoover: Produce more fossil fuels, open more public lands to exploitation, raise tariffs, limit immigration, allow segregation, minimize regulation, and reduce taxes. The government idealized the era when alternative gender identities were considered illnesses, women were expected to be homemakers, and white men dominated society. Republicans wanted not just to revive the old energy system but many other aspects of the past, including racial discrimination, nativism, sexism, religious fundamentalism, and homophobia, none of which had assisted the great energy transition or improved the standard of living.

During that transition, the cities challenged and gradually subordinated rural America, and new corporations displaced the railroads to dominate the economy and become the basis of the Dow Jones Industrial Average. A century later, however, those industries—notably oil, gas, and the automotive companies—were losing their dominance to digital corporations and renewable energies. These shifts had political consequences. The coal, oil, and gas industries supported politicians who opposed alternative energies. The industrial heartland of the old energy regime was struggling, and it began to vote Republican in search of its lost prosperity. The decline of blue-collar jobs and the rise of digital industries created anxiety among workers, and they responded to populist candidates. Something similar had happened during the turbulent 1890s. When the charismatic William Jennings Bryan ran for president in 1896, he won twenty-two states, most of which Donald Trump also carried in 2016 and 2024. Both were strong in the South, Midwest, and Mountain West, especially in rural areas. Bryan rejected evolutionary theory, and his supporters were the spiritual ancestors of Republican evangelical voters. Bryan did not criticize Jim Crow laws; Trump denounced Black Lives Matter. Both men were suspicious of foreign alliances. Bryan resigned as Wilson's secretary of state because he favored neutrality in World War I. Trump told supporters their taxes should solve domestic problems not pay for foreign wars. Bryan railed against the gold standard and advocated the free coinage of silver. He declaimed, "You shall not crucify mankind upon a cross of gold."[1] Trump also supported an alternative currency. He declared, "The United States will be the crypto capital of the planet and the Bitcoin superpower of the world." He launched his own cryptocurrency company, World Liberty Financial.[2]

The great energy transition once had made possible the engineering of nature, which Trump expected to continue by pumping in cryptocurrency and more fossil fuels. He claimed this would "usher in a golden age of American energy dominance and the extraordinary financial and security benefits that it brings to our citizens."[3] Trump linked energy with national renewal and thought, the more the better.[4] In his first term, he invested in nuclear power, exported coal, increased fracking for oil and gas, supported a new pipeline to Mexico, and opened federal lands to drilling and mining. In his second term, he mocked renewables and kept repeating, "Drill, baby, drill." His administration defunded research on climate change, hindered adoption of green energies, and loosened pollution standards. He withdrew from the Paris Climate Accords, and he proposed to end subsidies for electric vehicles and to halt what he called "the war on oil and natural gas."[5]

Yet his attempts to preserve the old energy regime were misguided. For by 2017 both scientists and investors could see that fossil fuels were not infinite resources, that climate change was real, and that alternative energies had become decisively less expensive than coal or oil. Coal power plants were closing because they were more expensive than solar and wind power. Burning fossil fuels guaranteed not prosperity but pollution, species extinction, wildfires, and climate emergencies.[6] A century earlier, Americans had no inkling that intensified use of fossil fuels would lead to global warming. With an inadequate grasp of energy, entropy, or technological change, they failed to comprehend the effects of their energy transition. In 2024, American voters had access to better concepts and comprehensive information, yet half of them still thought about energy as they had for the last hundred years. They elected a government that promised to turn back the clock and preserve the high polluting, fossil fuel, energy regime.

Halting the alternative energy transition might have been possible decades ago. But by 2024 in many cities and states alternative energies had achieved technological momentum. Salt Lake City, St. Paul, Minnesota, and Boulder, Colorado, expected to be using 100 percent renewable energy by 2030. Many communities already have achieved that goal, including Burlington, Vermont, and Denton, Texas. The state of Iowa produced more than 60 percent of its electricity with windmills, and Colorado planned to be based 100 percent on renewables by 2030. But other states remained locked in the old energy system. As late as 2023 Ohio and Indiana produced more than 80 percent of their electricity from coal and natural gas, with negligible wind or solar installations. In contrast, renewable energies were as well-established in Republican strongholds of Texas, Oklahoma, Arizona, Wyoming, and South Dakota as they were in Democratic California, Vermont, Maine, and Minnesota. Abroad, renewable energies supplied a majority of the electricity in many countries.[7] In short, the second Trump administration could slow down but it could not stop the transition. As had been the case in the United States between 1875 and 1929, the transition was driven sporadically by the federal government but persistently by cities, states, and private enterprise. The former Confederate states such as Alabama, Louisiana, and South Carolina had been slow to electrify or to adopt automobiles before 1920, and again they were slow to change.[8] But most of the nation was in transition by March 2025, when for the first time in a century, fossil fuels produced less than half of the nation's electricity.[9]

However, installing alternative energies did not lead to reduced consumption. For more than a century, the United States had grown accustomed to rising levels of energy use. By 2024, its energy consumption per capita was double that of Germany, France, Japan, or Switzerland.[10] Indoor plumbing, heating, and air conditioning, once luxuries, had become necessities. Cars had grown larger, and ownership increased. Houses were better insulated, but they also grew larger. Appliances became more efficient, but Americans bought more of them. Muscles no longer powered lawn mowers, drills, and other tools. Television screens and refrigerators expanded. Computer servers and networks required ever more power. Adopting alternate energies by itself did not restrain the voracious levels of consumption that had been unleashed and naturalized. Even if alternative technologies were adequate to supply the entire United States, the spendthrift habits fostered during the great energy transition would remain.

ACKNOWLEDGMENTS

The Wilson Library and Walter Library at the University of Minnesota provided indispensable resources, as did the Boston Public Library, the libraries at MIT, and the library of the University of Southern Denmark. I also drew upon the HathiTrust, Internet Archive, and Library of Congress. This study draws on a lifetime of research, including fruitful contact with Leo Marx, Merritt Roe Smith, Thomas Hughes, Wiebe Bijker, Ronald Martin, Martin Melosi, Tom Misa, Richard Hirsh, Mick Gidley, Jeffrey Meikle, Klaus Benesch, Robert Rydell, Rob Kroes, Thomas Zeller, Christopher Jones, Sandy Isenstadt, and Rosalind Williams. My understanding of energy transitions was sharpened in an online seminar on energy history organized in 2020 by Bernie Carlson and Erik Conway. In February 2023 I presented an overview of this book to colleagues at the University of Southern Denmark and benefited from their comments. Miles Orvell read an early draft of the manuscript, and Julie Cohn made useful comments on energy transitions and convinced me to add chapter 3. In 2024, two anonymous peer reviewers suggested useful refinements and perspectives. For four decades, it has been my good fortune to work with the excellent staff at the MIT Press, and I thank them once again. None of these fine people are responsible for any faults that may remain.

If it seems that I have slighted a topic, it may be discussed in greater depth in one of my previous books. A collective index of their contents can be found at https://conservancy.umn.edu/items/235ad6fd-3401-4598-9979-7b04561ab872.

NOTES

INTRODUCTION

1. Føllesdal, "The *Lebenswelt* in Husserl"; Ingram, *Habermas*.

2. Habermas, *Theory of Communicative Action*, vol. 2, 124. I do not attempt in this book to synthesize theories of the social construction of technology with the extensive writings of Husserl, Habermas, and other philosophers. Such a project would interest a subgroup of specialists but not the more general audience for whom this book is intended.

3. Schatzberg, *Technology*, 11; Marx, "Technology: The Emergence of a Hazardous Concept," 967–968.

4. Schatzberg, *Technology*, 72. On the construction of technological systems, see Bijker, Hughes, and Pinch, eds., *The Social Construction of Technological Systems*, and Bijker and Law, *Shaping Technology/ Building Society*.

5. Martin, *American Literature and the Universe of Force*, xi.

6. Martin, *American Literature and the Universe of Force*, 32–58, 93.

7. Thurston, "The Trend of National Progress," 306.

8. See Nye, "Progress or Entropy?," chapter 11 in *America as Second Creation*, 261–282.

9. Nye, *Technology Matters*, 17–66.

10. Gordon, *The Rise and Fall of American Growth*, 66–67.

11. Gordon, *The Rise and Fall of American Growth*, 527.

12. Gordon, *The Rise and Fall of American Growth*, 320–322.

13. Nye, *Consuming Power*, 251–254.

14. Smil, *Energy Transitions*, 224.

15. Sovacool, "How Long Will It Take?," 202–207.

16. Hughes, "Afterword," 174.

17. Nye, *Electrifying America*, 261, 187; Nye, *America's Assembly Line*, 32.

18. Hughes, *Networks of Power*, 14–17.

19. Hughes, "Technological Momentum," 111; Hughes, *American Genesis*, 460.

CHAPTER 1

1. Brewster, *A Treatise on the Kaleidoscope*. In the 1870s, Charles Busch, a Prussian immigrant to Massachusetts, received patents for improvements in kaleidoscope design. His instruments were the best available in the United States, and they adorned many parlors.

2. Brewster, *A Treatise on the Kaleidoscope*, 75.

3. Groth, "Kaleidoscopic Vision and Literary Invention in an 'Age of Things,'" 217–218.

4. A survey of US publications in the Hathai Trust between 1860 and 1930 reveals the frequency of use of "kaleidoscope" and "kaleidoscopic":

1860s	2,714
1870s	4,655
1880s	5,887
1890s	8,695
1900s	10,729
1910s	11,338
1920s	8,169
1930s	6,642
1940s	6,407

5. "Vers de Societé," *San Francisco Daily Evening Bulletin*, January 9, 1875, 2.

6. Coolidge, "1927 Budget Message," December 5, 1927, https://www.presidency.ucsb.edu/documents/1927-budget-message.

7. See Economic History Association, "Hours of Work in U.S. History," https://eh.net/encyclopedia/hours-of-work-in-u-s-history/.

8. Rodgers, *The Work Ethic in Industrial America*, 156–175.

9. Cited in Leiserson, *Adjusting Immigrant and Industry*, 127.

10. Nasaw, *Going Out*, 5.

11. Goldfield and Brownell, *Urban America*, 244–245.

12. Taylor, *Between the Gates*, 73–74.

13. "The City of Pittsburgh," 68.

14. Babcock, *Our American Resorts*, 10.

15. Aron, *Working at Play*, 138–143.

16. Silber, *The Romance of Reunion*, 69, 70.

17. Oettermann, *The Panorama*, 49–143.

18. Hales, *Silver Cities*, 69–71.

19. Trachtenberg, *Reading American Photographs*, 175.

20. Miller, *Picture Postcards in the United States*.

21. Cox, "Regionalism," 687.

22. Dewey, "Americanism and Localism," 686.

23. Kaplan, "Nation, Region, Empire," 250–251.

24. Sharp, *City Life and Its Amelioration*, 31–32.

25. McCabe, *New York by Gaslight*, 144.

26. *New York Illustrated*, 16.

27. Hoover, "Address to the Gridiron Club," December 14, 1929, https://www.presidency.ucsb.edu/documents/address-the-gridiron-club-3.

28. Bowen, "An Appeal to the King," 26.

29. Bowen, "An Appeal to the King," 30.

30. *Plessy v. Ferguson*, 163 U.S. 537 (1896). By a 7 to 1 vote, the Supreme Court ruled that the "separate but equal" laws and practices of state governments were constitutional.

31. Kemble, *Further Records, 1848–1883*, 145.

32. Martin, *A Vocabulary Study of "The Gilded Age"*; Twain and Warner, *The Gilded Age*, chapter 47.

33. Paine, *Mark Twain, A Biography*, vol. 2, 1343.

34. Cather, "Amusements," 6.

35. Register, *The Kid of Coney Island*, 175–177.

36. "Something New on Keith Program," *Charlotte Observer*, March 26, 1922.

37. Emerson, "Quotation and Originality," 179.

38. On Henry Adams, see Nye, *America as Second Creation*, 274–279.

39. Scranton and Fridenson, *Reimagining Business History*, 23.

40. Gordon, *The Rise and Fall of American Growth*, 13–16.

CHAPTER 2

1. Smil, *Energy in World History*, 388.

2. Gordon, *The Rise and Fall of American Growth*, 163.

3. Nye, *Consuming Power*, 122–123

4. Nye, *Consuming Power*, 251–254.

5. Smil, *Energy Transitions*, 224.

6. For more analysis, see Nye, "A Model for Heterogeneous Energy Transitions."

7. Nye, *Electrifying America.*

8. Wattenberg, *The Statistical History of the United States*, 819.

9. Jones, *Routes of Power* 140–144.

10. Rosenberg and Trajtenberg, "A General-Purpose Technology at Work," 63, 72.

11. Census Bureau, *Historical Statistics of the United States, Colonial Times to 1970*, Series A 57–72, "Population in Urban and Rural Territory, by Size of Place: 1790 to 1970," 11.

12. Nye, *American Illuminations*, 85–108.

13. Nye, *American Illuminations*, 56–59.

14. Nye, *American Illuminations*, 53.

15. Nye, *American Illuminations*, 161–170.

16. Danbom, *Born in the Country*, 187.

17. Edwards, *Comparative Occupation Statistics for the United States*, 101.

18. Rabenek, "Edison and Modern Construction," 471.

19. Bix, *Inventing Ourselves Out of Jobs?*, 18–19.

20. Hunter, *Waterpower in the Century of the Steam Engine*, 494.

21. Nye, *Consuming Power*, 82–94.

22. Stilgoe, *Metropolitan Corridor*, 77, 81.

23. Stilgoe, *Metropolitan Corridor*, 102.

24. Krugman, "Superstar Cities," summarizes the argument of his "Increasing Returns and Economic Geography."

25. Nye, *Electrifying America*, 4–7, 29–33.

26. Nye, *Consuming Power*, 198.

27. Nye, *Electrifying America*, 245–286.

28. Krugman, "Superstar Cities."

29. Smith, *Harpers Ferry Armory and the New Technology.*

30. Hounshell, *From the American System to Mass Production.*

31. Nye, *America's Assembly Line*, 20–32.

32. Census Bureau, *Historical Statistics*, Series S 1–14, "Total Horsepower of All Prime Movers: 1849 to 1970."

33. Gordon, *The Rise and Fall of American Growth*, 269–270.

34. McCraw, *Prophet of Innovation*, 270.

35. Nichols, "Why Schumpeter Was Right," 1024.

36. For patent office statistics, see "U.S. Patent Activity Calendar Years 1790 to the Present," https://www.uspto.gov/web/offices/ac/ido/oeip/taf/h_counts.htm.

37. Thurston, "The Trend of National Progress," 310.

38. Census Bureau, *Historical Statistics*, Series W 30–54, "Indexes of Output per Man-Hour, Production Workers."

39. Rodgers, *The Work Ethic in Industrial America*, 169.

40. Nye, *America's Assembly Line*, 101.

41. Scranton, *Endless Novelty*, 99.

42. Bohlen, "Assumption of Risk."

43. Nye, *Consuming Power*, 86–88.

44. Nye, *Consuming Power*, 88–89.

45. Montgomery, *Workers Control in America* and *Citizen Worker*.

46. Sombart, *Why Is There No Socialism in the United States?*

47. Census Bureau, *Historical Statistics*, Series A 172–194, "Population of Regions."

48. Foner, "Why Is There No Socialism in the United States?," 76.

49. Bellamy, *Looking Backward*.

50. Lipow, *Authoritarian Socialism in America*.

51. Gillette, *The Human Drift*, v–vi.

52. Gillette, *The Human Drift*, 89, 92.

53. Gillette, *The Human Drift*, 108.

54. Howells, *A Traveler from Altruria*.

55. Bierce, "Ashes of the Beacon," 21–22.

56. Tocqueville, *Democracy in America*, vol. 2, 144–145.

57. Bryce, *The American Commonwealth*, vol. 2, 293.

58. McCullough, *Old Wheelways*, 205–222.

59. Harper, *Life and Work of Susan B. Anthony*, vol. 3, 1293.

60. Trescott, "The Bicycle, a Technical Precursor of the Automobile."

61. Nye, *Electrifying America*, 94.

62. Trachtenberg, *The Incorporation of America*, 125.

63. National Museum of American History, exhibit, "America on the Move," https://americanhistory.si.edu/explore/exhibitions/america-on-the-move.

64. Wosk, "The Impact of Technology on the Human Image in Art," 148.

65. Baldwin, "When Cities Treated Cars as Dangerous Intruders."

66. Gordon, *The Rise and Fall of American Growth*, 100.

67. Gordon, *The Rise and Fall of American Growth*, 321.

68. Gusfield, *Symbolic Crusade*.

CHAPTER 3

1. Reynolds, "The Education of Engineers in America Before the Morrill Act."

2. Friedel, *The Culture of Improvement*, 392–399.

3. Layton, *Revolt of the Engineers*; Noble, *America by Design*; and Sinclair, *Centennial History of the American Society of Mechanical Engineers*.

4. Nye, *Conflicted American Landscapes*, 9–28.

5. Nye, *American Technological Sublime*, 45–142.

6. Tichi, *Shifting Gears*, 97–99.

7. Richards, *The Man of Tomorrow*, 213.

8. Nye, *America as Second Creation*, 9–20, 290–292.

9. Nye, *America as Second Creation*, 109–116, 153–157, 206–211, and 284–292.

10. Newell, *Irrigation in the United States*, 393–394.

11. Smyth, *The Conquest of Arid America*, 329.

12. For a critique of this idea, see Nye, *Conflicted American Landscapes*, 4–6.

13. Nye, *Conflicted American Landscapes*, 70–74; Boime, *The Magisterial Gaze*, 140.

14. Nye, *Conflicted American Landscapes*, 93–94, 99.

15. Nye, *Seven Sublimes*, 17–34.

16. Nevins, *John D. Rockefeller*.

17. See "Republican Party Platform of 1908," The American Presidency Project, June 16, 1908, https://www.presidency.ucsb.edu/documents/republican-party-platform-1908.

18. Nye, *American Technological Sublime*, 53, 56, 60–61.

19. Cited in Nye, *America as Second Creation*, 194.

20. *Census of Manufacturers*, 1929, vol. 1, table 11, 223; vol. 2, 959.

21. US Energy Information Administration, "Production of Crude Oil," https://www.eia.gov/dnav/pet/hist/LeafHandler.ashx?n=pet&s=mcrfpus2&f=a.

22. Wattenberg, *The Statistical History of the United States*, 819.

23. National Research Council, *Assessing and Managing the Ecological Impacts of Paved Roads*, 37.

24. Seely, *Building the American Highway System*, 19–20. See also Meinig, *The Shaping of America*, 3–33.

25. Zeller, *Consuming Landscapes*, 59–65.

26. McClelland, *Building the National Parks*, 146–147.

27. Quoted by Zeller in *Consuming Landscapes*, 21.

28. Kuta, "The Man Who Invented Water Skiing."

29. Jackson, "The Abstract World of the Hod-Rodder," 146–147; Nye, *Narratives and Spaces*, 185–186.

30. Melosi, *Effluent America*, 55.

31. Melosi, *Effluent America*, 58.

32. Freund, *American Sunshine*, 51–53.

33. Melosi, *Effluent America*, 107.

34. Markowitz and Rosner, "The Street of Walking Death," 529.

35. Derickson, "On the Dump Heap."

36. Markowitz and Rosner, "The Street of Walking Death," 533.

37. Sheibach et al., "Controlling Acid Mine Drainage," 45.

38. Barringer, "Despite Cleanup at Mine, Dust and Fear Linger," 294.

39. Nye, *Electrifying America*, 204.

40. LeCaine, *Mass Destruction*, 174–176.

41. Danbom, *Born in the Country*, 131.

42. Mock, "Rainfall in the Garden of the United States Great Plains."

43. Opie, *Ogallala*, 59.

44. Raban, *Bad Land*.

45. Danbom, *Born in the Country*, 148.

46. Kline, *Consumers in the Country*, 65.

47. Kline, *Consumers in the Country*, 286.

48. Hirsh, *Powering American Farms*, 106–108.

49. Census Bureau, *Historical Statistics*, Series K 445–485, "Man Hours per Unit and Yield per Unit," 500.

50. Census Bureau, *Historical Statistics*, Series W 67–81, "Index of Farm Production," 953.

51. Census Bureau, *Historical Statistics*, Series K 1–16, "Farm Population," 457.

52. Berry, *The Unsettling of America*, 128.

53. Smil, *Energy in World History*, 182–183.

54. Crabb, *The Hybrid-Corn Makers.*

55. Berry, *The Unsettling of America*, 133.

56. Nye, *America as Second Creation*, 243–244.

57. Steinberg, *Down to Earth*, 175.

58. Cohn, *The Grid.*

59. Spence, *Dispossessing the Wilderness.*

60. Nixon, *Slow Violence and the Environmentalism of the Poor*, 154–165.

61. Mutel, *The Enchanted Horizon*, 69–70.

62. Nash, *Grand Canyon for Sale*, 46.

63. Steinberg, *Down to Earth*, 146.

64. Mutel, *The Enchanted Horizon*, 84–85.

65. Hornaday, "The Extermination of the American Bison," 376–377.

66. Lueck, "The Extermination and Conservation of the American Bison," S609–S616.

67. Marsh, *Man and Nature.*

68. Cited in Hays, *Conservation and the Gospel of Efficiency*, 124.

69. Hays, *Conservation and the Gospel of Efficiency*, 127.

70. Steinberg, *Down to Earth*, 141–143.

71. Nye, *Henry Ford*, 83.

72. Nye, *Henry Ford*, 81.

73. Ford and Crowther, *My Life and Work*, 280.

74. Nye, *American Illuminations*, 125–130.

75. "Niagara Falls from a New Point of View," *Scientific American*, 227. See also Nye, *American Technological Sublime*, 136–137.

76. Nye, *Electrifying America*, 58.

77. *New York Tribune*, September 5, 1907. See also Nye, *Electrifying America*, 58–60.

78. Jackson, *Pastoral and Monumental*, 19.

79. Marchand, *Advertising the American Dream*, 227. See also Marchand, *Creating the Corporate Soul.*

80. Nye, *American Technological Sublime*, 177–178.

81. Oppenheim, *The Olympian*, 2; Dreiser, *A Hoosier Holiday*, 180.

82. Coyne and Schiller, "John Sloan's Urban Encounters," 55–57.

83. Schleier, *The Skyscraper in American Art*, 53–54.

84. Williams, *Sour Grapes*, 78.

85. Tsujimoto, *Images of America*, plates 39, 40, 46, 47, 50, 54, 57.

86. Adams, *Our Business Civilization*, 253.

CHAPTER 4

1. Yost, *The Principles of Journalism*, 7, 8, 78.

2. "A Moving Picture on the Front Page," *Miami Herald*, December 21, 1922.

3. Funk, "Republicans' Views on Evolution."

4. Tattersall, "Thermal Degeneration," 17–34.

5. Nash, *The Nervous Generation*, 149.

6. Digital History, "Journalist H. L. Mencken's Account of the Scopes Trial," https://www.digitalhistory.uh.edu/disp_textbook.cfm?smtID=3&psid=1077.

7. Bryan College, "About," https://www.bryan.edu/about/

8. *McGuffey's Fifth Eclectic Reader*, text 49, "Behind Time," https://www.gutenberg.org/ebooks/15040.

9. McCrossen, *Marking Modern Times*, 189.

10. O'Malley, *Keeping Watch*, ix.

11. Adas, *Machines as the Measure of Men*, 259.

12. McCrossen, *Marking Modern Times*, 18.

13. McCrossen, *Marking Modern Times*, 138.

14. Thoreau, "Sounds," in *Walden*, 122.

15. Cited in Gleick, *Faster*, 44.

16. Valosik, "Synchronized Swimming Has a History"; Sydnor, "A History of Synchronized Swimming," 252–255.

17. Kracauer, *The Mass Ornament*, 77.

18. For newspaper coverage of the Johnstown Flood, see Chronicling America, Library of Congress, https://chroniclingamerica.loc.gov/search/pages/results/?state=&date1=1888&date2=1890&proxtext=Johnstown+Flood&x=18&y=12&dateFilterType=yearRange&rows=20&searchType=basic&sort=date. On June 1, the day after the disaster, it was reported in the following papers: *The Wheeling Daily Intelligencer*, *Omaha Daily Bee*, *Evening Star* (Washington DC), *Sacramento Daily Record-Union*, *The New York Evening World*, *The Lancaster Daily Intelligencer*, *The Helena Independent*, *The Pittsburg Dispatch*, *The Salt Lake City Herald*, *The Wichita Eagle*, *The St. Paul Daily Globe*, *The Los Angeles Daily Herald*, *The Portland Daily Press*, *The Cheyenne Daily Leader*, and many more.

19. Kern, *The Culture of Space and Time*, 34.

20. Adams, *The Degradation of the Democratic Dogma*, 274–276.

21. Wells, *The Time Machine*.

22. Quoted by Twain in *Roughing It*, 29.

23. Stover, *The Routledge Historical Atlas of the American Railroads*, 37–41.

24. Nye, *America's Assembly Line*, 44–45.

25. Missal, *Seaway to the Future*.

26. Nye, *Seven Sublimes*, 20–29.

27. Nye, *American Illuminations*, 64–78, 134–159; Nye, *Electrifying America*, 29–84.

28. Isenstadt, *Electric Light*, 165–168.

29. Cited in Nye, *Electrifying America*, 77.

30. Jaffe, *Joseph Stella's Symbolism*, comments on plate 10.

31. Marshall, *Seeing America*, 19.

32. Frank, *Our America*, 175.

33. Beard, *American Nervousness*, vi.

34. Beard, *American Nervousness*, 98–99.

35. Beard, *American Nervousness*, 103–104.

36. Beard, *American Nervousness*, 114.

37. "The Human Coalburner," 12, 37.

38. Rabenbach, *The Human Motor*, 116.

39. See Taylor, *The Principles of Scientific Management*, and Scranton, *Endless Novelty*.

40. Israel, *Edison*, 150–151.

41. Smith, *Vocal Tracks*, 82–83.

42. Nye, *American Illuminations*, 120.

43. "Centennial Letter," *New Hampshire Sentinel*, May 18, 1876.

44. "The Centennial Exhibition," *Baltimore Sun*, May 24, 1876.

45. Magill, "The Close of the Exposition," 757.

46. "Buffalo Bill's Review of the Representative Rough Riders of the World."

47. Nye, *Electrifying America*, 33–47, 367–379; Nye, *Narratives and Spaces*, 113–128; Nye, *American Illuminations*, 109–132, 176–182.

48. Williams, *Dream Worlds*, 84.

49. Nye, *Electrifying America*, 39; Cameron, *The World's Fair*, 275.

50. Reik noted in 1927: "An electric fountain, with kaleidoscopic color displays, has been constructed here especially to entertain Denver's citizens and guests at night, while they listen to the Municipal Band concerts" (Reik, *A Tour of America's National Parks*, 23).

51. "Display of Fireworks at the Columbian Exposition," 359.

52. Cameron, *The World's Fair*, 816.

53. Kasson, *Amusing the Million*, 31.

54. Marshall, *Seeing America*, 240.

55. Kasson, *Amusing the Million*, 72.

56. Register, *The Kid of Coney Island*, 118

57. Page, *The City's End*, 29–69.

58. Frank, *Coney Island*, 83.

59. US Patent 758341.

60. Kouwenhoven, *Half a Truth Is Better than None*, 120–124.

61. Cameron, *The World's Fair*, 670.

62. Nye, *American Technological Sublime*, 148.

63. Gunn, "Summer Resorts," 360.

64. Paine, "The New Coney Island," 533.

65. The film can be seen at Library of Congress, *Coney Island at Night*, https://www.loc.gov/item/mp73008500/.

66. Cited in Register, *The Kid of Coney Island*, 120–121.

67. Wohl, *A Passion for Wings*, 5–28.

68. Nye, *American Technological Sublime*, 202–203.

69. Haller, "Riding in a German Airship," 6.

70. Catherine Oglesby, "Consciousness of Battle with the Elements is Thrilling Part of Riding in Airplane," *Fort Worth Star-Telegram*, June 8, 1919, 41.

CHAPTER 5

1. Riis, *How the Other Half Lives*, 296.

2. For an overview, see Northern Illinois Digital Library, "American Populism, 1876–1896," https://digital.lib.niu.edu/illinois/gildedage/populism; Kazin, *The Populist Persuasion*.

3. Kolko, *The Triumph of Conservatism*, 79–189.

4. Tarbell, *The History of Standard Oil*.

5. Nye, *Image Worlds*, 135–144.

6. Kolko, *The Triumph of Conservatism*, 127–132, 165–167.

7. Kolko, *The Triumph of Conservatism*, 36.

8. Steffens, *The Shame of the Cities*, 135–140.

9. Steffens, *The Shame of the Cities*, 164–165.

10. Roosevelt, "The Present State of Civil Service Reform," 239–240.

11. Roosevelt, "The Present State of Civil Service Reform," 242–243.

12. Schlereth, *Victorian America*, 76.

13. Plunkitt, "The Curse of Civil Service Reform."

14. Sinclair, *The Jungle*.

15. Zinn, *A People's History of the United Sates*, 353.

16. See National Archives, "16th Amendment to the U.S. Constitution: Federal Income Tax (1913)," https://www.archives.gov/milestone-documents/16th-amendment.

17. Wilson, *The City Beautiful Movement*; Nye, *American Illuminations*, 158–183.

18. Aldrich, "Unguarded Gates," 57.

19. Garcia, "A History of Mexicans and Mexican Americans."

20. Hirschman and Mogford, in "Immigration and the American Industrial Revolution," table 1.

21. Grant, *The Passing of the Great Race*, 263.

22. Fitzgerald, *The Great Gatsby*, 17.

23. Daniels, *Asian America*, 123–125.

24. Øverland, *Immigrant Minds*.

25. Øverland, *Immigrant Minds*, 94–95.

26. Øverland, *Immigrant Minds*, 144–145.

27. Selig, *The Cultural Gifts Movement*, 8.

28. Chauncey, *Gay New York*.

29. "Current Magazines," *New York Times*, May 30, 1926, B19.

30. Levine, *Eugenics*.

31. Wright, *Moral Energy in America*, 70–73.

32. Rydell, *All the World's a Fair*, 64.

33. Rydell, *All the World's a Fair*, 64–65.

34. Rydell, *All the World's a Fair*, 131.

35. Strong, *Our Country*, chapter 13.

36. Lorimer, "Courage and Common Sense," 20.

37. Hochschild, *American Midnight*, 175–178, 144–148.

38. Smith, *Age of Fear*, 157–162, 169.

39. Palmer, "The Case Against the Reds," 63–64.

40. Goldberg, *Discontented America*, 92.

41. Johnson, "The Riots," 242.

42. Hochschild, *American Midnight*, 251–254.

43. Hirsch, *Riot and Remembrance*.

44. NAACP, *Thirty Years of Lynching in the United States, 1889–1918*. The supplement for later years is at https://collections.library.yale.edu/catalog/2077126.

45. Goldberg, *Discontented America*, 117–139.

46. Cited in Nye, *Image Worlds*, 79.

47. Hughes, *The Big Sea*, https://gutenberg.ca/ebooks/hughesl-bigsea/hughesl-bigsea-00-h-dir/hughesl-bigsea-00-h.html.

48. Shindo, *1927 and the Rise of Modern America*, 79.

49. Taylor, "W. E. B. Du Bois's Challenge to Scientific Racism," 449.

50. Lieberman, *Power Lines*, 176–177.

51. Enck, "Tuskegee Institute and Northern White Philanthropy."

52. See National Park Service, Booker T. Washington National Monument Virginia, "The Atlanta Exposition Address Part 2," https://www.nps.gov/bowa/learn/historyculture/atlanta2-1.htm.

53. See Smithsonian National Museum of American History, "Charles Hamilton Houston," https://americanhistory.si.edu/brown/history/3-organized/charles-houston.html.

54. Du Bois, *The Souls of Black Folk*, 18.

55. Du Bois, *The Souls of Black Folk*, 202.

56. Ritchie, *Major Taylor*.

57. Ritchie, *Major Taylor*, 90.

58. Peterson, *Only the Ball Was White*, 80–84, 207–215.

59. Hobbs, *A Chosen Exile*.

60. Walter Lippmann, "Introduction," in Sandberg, *The Chicago Race Riots*, vi.

61. See US History.org, "The Pledge of Allegiance," https://www.ushistory.org/documents/pledge.htm.

62. Pivar, *Purity Crusade*.

63. Donovan, "Framing and Strategy."

64. See Lesley Kennedy, "Why the 19th Amendment Did Not Guarantee All Women the Right to Vote," History.com, March 1, 2021, https://www.history.com/news/19th-amendment-voter-suppression.

CHAPTER 6

1. Tierney, "What Is Nostalgia Good For?"

2. Meyer, *Myths in Stone*, 155–156.

3. "Dedicated! The Act Which Completes the Washington Monument," 2.

4. "Dedicated! The Act Which Completes the Washington Monument," 3.

5. Hargrove, "Unveiling the Colossus," 200.

6. Blumberg, "A National Monument Emerges," 210–213.

7. See Andrew Boyd, "Miss Columbia," University of Houston, The Engines of Our Ingenuity, https://www.uh.edu/engines/epi2756.htm.

8. Nye, *American Technological Sublime*, 264–266.

9. Luckiesh, *Artificial Light*, 304.

10. Wilson Feted," *Morning Oregonian*, November 30, 1916, 2.

11. Blumberg, "A National Monument Emerges," 211–212; illustrations, 245, 247.

12. Nye, *Electrifying America*, 61.

13. Foner, *The Second Founding*, xxii.

14. Hofstadter, "Ulrich B. Phillips and the Plantation Legend," 122–123.

15. See Judson MacLaury, "The Federal Government and Negro Workers Under President Woodrow Wilson," US Department of Labor, March 16, 2000, https://www.dol.gov/general/aboutdol/history/shfgpr00.

16. Meyer, *Myths in Stone*, 217.

17. Kammen, *In the Past Lane*, 205.

18. Gatewood and Cameron, "Battlefield Pilgrims at Gettysburg National Military Park."

19. Linenthal, *Sacred Ground*, 93–97.

20. Wilson, "Address at Gettysburg," The American Presidency Project, July 4, 1913, https://www.presidency.ucsb.edu/documents/address-gettysburg-0.

21. Lears, *No Place of Grace*, xi.

22. On Adams and the dynamo, see Nye, *America as Second Creation*, 266–282.

23. Cram, *Walled Towns*, 16, 23; Lears, *No Place of Grace*, 203–207.

24. Cram, *Walled Towns*, 43.

25. Cram, *Walled Towns*, 46.

26. James, *Hawthorne*, 43–44.

27. Eliot, *The Waste Land*, 14–15.

28. Eliot, *Ash Wednesday*.

29. Webb, "A Pictorial Myth in the Pages of *Life*," 35–36.

30. Cited in Orvell, *The Death and Life of Main Street*, 16.

31. Bourne, "The Social Order in an American Town," 227–228.

32. Marchand, *Advertising the American Dream*, 260.

33. Orvell, *The Death and Life of Main Street*, 26.

34. Frost, "Preface."

35. Sears, *Sacred Places*, 3–11.

36. Runte, *National Parks*, 11–12.

37. Anderson, "The Kiss of Enterprise," 243–244; Nye, *Conflicted American Landscapes*, 93–98, 108; Nye, *American Technological Sublime*, 76. In 1921 a Union Pacific Railroad book on Yellowstone Park praised the Great Falls, which "leaps with the roar of a thousand dynamos 308 feet to the floor of the kaleidoscopic canyon, losing itself in a whirling nebula of spray which sends graceful streamers of mist incessantly upward" (14).

38. Ferris, *Our Native Land*, 360.

39. National Park Service, "Many Colored Canyon," n.p.

40. Kyle, *Yosemite, the World's Wonderland*, 31.

41. Van Slyck, *A Manufactured Wilderness*.

42. Roosevelt, *The Winning of the West*.

43. Nye, *America as Second Creation*, 82–85.

44. Hyde, *An American Vision*, 255–258.

45. Smith, "The Survey's Contributions to the National Parks Movement," 322.

46. Jakle, *The Tourist*.

CHAPTER 7

1. For an overview, see Paul, *The Myths That Made America*, 367–420, and Cawelti, *Apostles of the Self-Made Man*.

2. Tocqueville, *Democracy in America*, vol. 2, 104–105.

3. Nye, *America as Second Creation*, 40–43, 59–67, 91–93, 165–166, 283–284.

4. Turner, *The Frontier in American History*, 30, 203–204, 321.

5. "The President's Message," 63. Similar views are expressed on 65, 415, 421, 607, and 608.

6. Burch and Patterson, *Problems of American Democracy*, 84.

7. Wright, *Moral Energy in America*, 40, 46.

8. Wister, *The Virginian*; Grey, *Riders of the Purple Sage*; Wright, *The Winning of Barbara Worth*.

9. Nye, *Narratives and Spaces*, 27–29, 38–39.

10. Conwell, *Acres of Diamonds*, 18, 44.

11. Nash, *The Nervous Generation*, 127.

12. Strong, *Our Country*, chapter 13.

13. Lears, *Rebirth of a Nation*, 29.

14. Carnegie, "The Gospel of Wealth.," 658–650.

15. Carnegie, "The Gospel of Wealth," 664.

16. Hoover, "The Future of American Individualism," Hoover Institution, https://www.hoover.org/research/future-american-individualism.

17. Meyer, *The Positive Thinkers*, 32–45, 197.

18. Interview with the *Detroit News*, February 2, 1926, cited in Nye, *Henry Ford: Ignorant Idealist*, 113.

19. Trine and Ford, *The Power That Wins*, 52.

20. Smith, *Sunshine and Shadow in New York*, 29.

21. Smith, *Sunshine and Shadow in New York*, 72.

22. Smith, *Sunshine and Shadow in New York*, 244.

23. Martin, *The Secrets of the Great City*, 55.

24. "The Pace That Kills," *Fort Worth Star-Telegram*, September 29, 1903.

25. Bingham, "Bibliography of American Editions of Robinson Crusoe to 1830."

26. Skrabec, *William McKinley*, 41.

27. Thayer, *From Log Cabin to White House*, 138–139.

28. Ford, *Female Robinson Crusoe*, 76–77.

29. See the Encyclopedia of Science Fiction, "robinsonade," https://sf-encyclopedia.com/entry/robinsonade.

30. See Palmer, *Castaway Tales*.

31. Frazer, *Perseverance Island*, 1.

32. Selcraig, "The Real Robinson Crusoe."

33. In *Capital*, Marx argued that "Nature itself compels [Crusoe] to divide his time with precision between his different functions." But few of Crusoe's actions were forced upon him by nature. He begins with metal tools and weapons, and he did not need to invent the idea of domesticating animals, of agriculture, of tool making, or of a table, a chair, a fortified shelter, fishing, taming goats, weaving baskets, or making pots. After a short period of hardship, he had surplus food and free time. Marx, *Capital*, vol. 1, 47–49. See also Nail, *Marx in Motion*.

34. Samuelson, "An Exact Consumption-Loan Model of Interest."

35. White, "Reading and Rewriting."

36. For further critique, see Grapard and Hewitson, *Robinson Crusoe's Economic Man*, and Grapard, "Robinson Crusoe."

37. Defoe, *Robinson Crusoe*, 127.

38. Defoe, *Robinson Crusoe*, 187.

39. Defoe, *Robinson Crusoe*, 127.

40. Defoe, *Robinson Crusoe*, 127.

41. Some retreated to the countryside to practice simple living, including Helen and Scott Nearing, who left New York City to take up subsistence farming in the 1920s. Others who rejected technological abundance are discussed in Shi's *The Simple Life*.

42. Defoe, "An Essay upon Projects" and *A General History of Discoveries and Improvements*. In Defoe's sequel, Crusoe returns with livestock and settlers, and the island becomes part of the Atlantic economy, making it vulnerable to attack.

43. Adas, *Machines as the Measure of Men*.

44. Twain, *A Connecticut Yankee in King Arthur's Court*, chapter 7.

45. Oldenziel, *Making Technology Masculine*, 25–42. See also Lerman, "Industrial Genders, Constructing Boundaries."

46. Lemelson-MIT, "Josephine Cochrane," https://lemelson.mit.edu/resources/josephine-cochrane.

47. Iowa State University Institute for Transportation, "Maria Beasley: Engineering Dynamo," https://intrans.iastate.edu/news/maria-beasley-engineering-dynamo/.

48. Vella, *George Washington Carver*.

49. Nixon, "Special Message to the Congress Proposing Patent Modernization and Reform Legislation."

50. Encyclopedia of Science Fiction, "edisonade," https://sf-encyclopedia.com/entry/edisonade.

51. Cited by Edison's assistant, Jehl, in *Menlo Park Reminiscences*, vol. 1, 160, 169.

52. On doubts that Edison could subdivide the electric current, see Bazerman, *The Languages of Edison's Light*, 186.

53. "Edison in His Workshop," 607.

54. "Edison's Light, Sketch of Curious Visitors at Menlo Park," *New York Herald*, December 29, 1879, 2.

55. In the 1990s Newt Gingrich celebrated "the spirit of free enterprise" and "the spirit of invention and discovery," and declared, "We have no caste system, no class requirements, no regulated professions, no barriers to entry." He saw Edison and Bill Gates as ideal citizens. Gingrich, *To Renew America*, 41–43.

56. "A Food Creator: Edison Invents a Machine That Will Feed the Human Race."

57. McClure, *Edison and His Inventions*, 13.

58. "The Great Edison Scare," *Saturday Review*, January 10, 1880, 41.

59. Fiske, *Off-Hand Portraits of Prominent New Yorkers*, 113.

60. Dyer and Martin, with Meadowcroft, *Edison*.

61. See *The Guardian*, "Thomas Edison on the Progress of Electricity—Archive, 1896," November 25, 2016, https://www.theguardian.com/science/2016/nov/25/electicity-thomas-edison-renewable-energy-interview-1896.

62. Bok, *The Americanization of Edward Bok*.

CHAPTER 8

1. See Library of Congress, "Dow Jones Industrial Average First Published," https://guides.loc.gov/this-month-in-business-history/may/djia-first-published.

2. Stuart A. Thompson and William Power, "The Ins and Outs of the Dow Jones Industrial Average," *Wall Street Journal*, January 25, 2017, https://www.wsj.com/graphics/djia-components-history/?gaa_at=eafs&gaa_n=ASWzDAiR-1epxLLCkF8j7HdPfUjnSpSMctrHVVyC5lBTv-X16MlgwJzlC9aG&gaa_ts=683ec85e&gaa_sig=5Bw7SyWHkEwJRzJfz4VbbhdgkxLCYa_0RpSGt4ZMN_mncCe0d4KDli6m5AdrPcS6N6JJz-9MTDteUAXz_HIcqw%3D%3D. Corporations included in the Dow on April 1, 1900, were Amalgamated Copper, American Smelting & Refining, American Sugar, Continental Tobacco, International Paper, National Lead, Peoples Gas, Tennessee Coal & Iron, U.S. Leather, U.S. Rubber, and U.S. Steel.

3. Thompson and Power, "The Ins and Outs of the Dow Jones Industrial Average." Resource extraction and heavy industry: Allied Chemical, American Can, American Smelting, American Sugar, American Tobacco B, Bethlehem Steel, International Nickel, Texas Gulf Sulphur, U.S. Steel, Union Carbide. Energy transition companies: Atlantic Refining, Chrysler, Curtis-Wright Aeronautical, General Electric, General Motors, General Railway Signal, Goodrich, International

Harvester, Mack Truck, Nash Motors, Standard Oil of New Jersey, Texas Company [oil], Westinghouse Electric. Consumer goods and media: General Foods; Johns-Manville; National Cash Register; Paramount; Radio Corporation; Sears, Roebuck; Woolworth.

4. Nye, *America's Assembly Line*, 53.

5. See United States Mint, "The History of U.S. Circulating Coins," https://www.usmint.gov/learn/history/us-circulating-coins.

6. See the Bureau of Engraving and Printing, "History," https://www.bep.gov/currency/history.

7. Boorstin, *The Americans*, 146.

8. Mendenhall, *American Trademarks*.

9. On the Great White Way, see Nye, *Electrifying America*, 51–61.

10. On electric signs, see Nye, *American Technological Sublime*, 173–198; Nye, *American Illuminations*, 139–153, 197–202.

11. See Louisville Slugger Museum & Factory, "From Butter Churns to Baseball Bats: History in the Making," https://www.sluggermuseum.com/about-us/our-history.

12. Dreiser, *Sister Carrie*, 145.

13. Census Bureau, *Historical Statistics*, Series A 143–157, "Median Age of the Population, by Race, Sex, and Nativity: 1790 to 1970."

14. See Siddhartha Sanghi and Amy Smaldone, "The Evolution of the Racial Gap in U.S. Life Expectancy," Federal Reserve Bank of St. Louis, January 27, 2022, https://www.stlouisfed.org/on-the-economy/2022/january/evolution-racial-gap-us-life-expectancy.

15. Schlereth, *Victorian America*, 141.

16. Meikle, *Twentieth Century Limited*.

17. Orvell, *The Real Thing*, 144–145.

18. Watney, "Making Strange," 173.

19. Nye, *Image Worlds*, 115–123.

20. Marchand, *Advertising the American Dream*, 165–166.

21. Lears, *Fables of Abundance*, 385.

22. Marchand, *Advertising the American Dream*, 77–80.

23. Marchand, *Advertising the American Dream*, 87.

24. Marchand, *Advertising the American Dream*, 115.

25. Marchand, *Advertising the American Dream*, 166. These were derived from the *tableaux vivant* (or living pictures), "a genre of theater entertainment that enjoyed

a moment of popularity toward the end of the nineteenth century." Families and friends used costumes and props to recreate a famous painting, historical scene, or biblical story.

26. Marchand, *Advertising the American Dream*, 188–189.

27. Cowan, *More Work for Mother*.

28. Nye, *Image Worlds*, 125–127.

29. Nye, *Image Worlds*, 126.

30. Boorstin, *The Americans*, 127.

31. Sears, Roebuck Corporation, *1902 Edition of the Sears, Roebuck Catalogue.*

32. Both developments relate to the maximization of profits through economies of scale and speed of throughput. Chandler, *The Visible Hand*, 224–235.

33. Beniger, *The Control Revolution*, 331–334.

34. Bates, *The Art and Literature of Business*, 146–147.

35. Israel, *Edison*, 297–301.

36. Nasaw, *Going Out*, 154–155.

37. Nasaw, *Going Out*, 156–157.

38. Nasaw, *Going Out*, 157.

39. "Amusements for Park Resorts," 485.

40. In circa 1910 the popular "erector set" consisted of hundreds of small parts that a child could assemble into bridges, buildings, or small machines. Again, transparency was essential.

41. Whitaker, "Before Horn & Hardart."

42. Orvell, *The Real Thing*, 145.

43. Ewen, *Captains of Consciousness*, 1–26.

44. Such an argument appears in Aronowitz, *False Promises*.

45. To both radicals and conservatives, radio, film, and mass-market leisure spelled the death of traditional community. T. S. Eliot lamented that the working classes had lost "traditional culture" and become listless due to the introduction of cinema, phonographs, and the automobile. The Frankfurt School saw the production of popular culture by movie studios, record companies, and radio stations as a threat. Eliot, *Notes Toward the Definition of Culture*; Jay, *The Dialectical Imagination*.

46. Jameson, "Reification and Utopia in Mass Culture."

47. Leach, *Land of Desire*, 124–150.

48. Douglas and Isherwood, *The World of Goods*, 65.

49. The culture of consumption was driven by emulative spending and the desire for social distinction. In England, by 1750, "What men and women had once

hoped to inherit from their parents, they now expected to buy for themselves. What were once bought at the dictate of need, were now bought at the dictate of fashion. What were once bought for life, might now be bought several times over. . . . Even 'necessities' underwent a dramatic metamorphosis in style, variety, and availability." McKendrick, Plumb, and Brewer, *The Birth of a Consumer Society*, 1. See also McCracken, *Culture and Consumption*, 3–31.

50. Mitchell, *Iconology*, 196.

51. Ihde, *Technology and the Lifeworld*, 74.

52. Schudson, *Advertising, the Uneasy Persuasion*.

53. Blaszczyk, *Imagining Consumers*, 93.

54. Blaszczyk, *Imagining Consumers*, 229.

55. Nye, *Henry Ford*, 38–39.

56. Benjamin, "The Work of Art in the Age of Mechanical Reproduction," 217–223.

57. See Sklar, *The Corporate Reconstruction of American Capitalism*; Cullis, "The Limits of Progressivism"; S. Leccese, "John D. Rockefeller, Standard Oil, and the Rise of Corporate Public Relations."

58. Hayes, "The Employees' Publication," 83.

59. Nye, *Image Worlds*, 61–70.

60. Hayes, "The Employees' Publication," 81–94.

61. Hayes, "The Employees' Publication," 85–86.

62. Hayes, "The Employees' Publication," 89.

63. O'Shea, "The Shop Paper as an Aid to Morale," 791–792.

64. Nye, *American Illuminations*, 140.

65. Chandler, *The Visible Hand*, 486.

66. Marchard, *Creating the Corporate Soul*, 164–166.

67. Lynd and Lynd, *Middletown*, 88.

68. Lynd and Lynd, *Middletown*, 83.

69. Lynd and Lynd, *Middletown*, 81.

CHAPTER 9

1. Shaw, *Erie Water West*, 186.

2. *Utica Sentinel*, November 8, 1825. See also "Celebration of the Completion of the Erie Canal," *Albany Daily Advertiser*, October 11, 1825.

3. Czitrom, *Media and the American Mind*, 13.

4. Mann, "Telegraphing of Election Returns."

5. See Ward, "The Meaning of Lindbergh's Flight."

6. The following draws on Bain, *Empire Express*, 645–669, and White, *Railroaded*, 17–39.

7. Nye, *American Technological Sublime*, 73–76.

8. White, *Railroaded*, 186–188.

9. Dayan, "Beyond Media Events," 392–393.

10. Taylor, *The Transportation Revolution*, 152.

11. Robertson, *America's Business*, 114.

12. Cronon, *Nature's Metropolis*, 121.

13. Platt, *The Electric City*, 18.

14. Jerome Redding Company, *Catalog*, 4–7.

15. Blonheim, *News over the Wires*, 48.

16. Blonheim, *News over the Wires*, 49–51.

17. Schiller, *Objectivity and the News*, 3–7.

18. Trachtenberg, *The Incorporation of America*, 38.

19. Czitrom, *Media and the American Mind*, 10.

20. Fischer, *American Calling*, 2.

21. Frost, "Radio Dreams That Can Come True," 18.

22. Douglas, *Inventing American Broadcasting*, xv.

23. RCA executive David Sarnoff declared: "When television has fulfilled its ultimate destiny, man's sense of physical limitation will be swept away, and his boundaries of sight and hearing will be the limits of the earth itself. With this may come a new horizon, a new philosophy, a new sense of freedom, and greatest of all, perhaps, a finer and broader understanding between all the peoples of the world." Horwitz, *The Irony of Regulatory Reform*, 178.

24. In the 1990s many thought the internet would foster a freer market and universal understanding. This "Third Wave" would break up big business and big government and promote individual autonomy. Gingrich, *To Renew America*, 7, 57.

25. Nye, *American Illuminations*, 197.

26. "Lights to Spell out High Ideals on 4th," *New York Times*, July 2, 1916, S4.

27. Sterling and Kittross, *Stay Tuned*, 510.

28. Wallace, *Calvin Coolidge*, 24.

29. Wallace, *Calvin Coolidge*, 2.

30. Wallace, *Calvin Coolidge*, 22.

31. Ward, "The Meaning of Lindbergh's Flight," 4–14.

32. Wallace, *Calvin Coolidge*, 21.

33. "Thomas A. Edison, Guest of the Institute," 151–152. See also David and Julia Bart, "IEEE Edison Medal," Engineering and Technology History Wiki, 2010, https://ethw.org/IEEE_Edison_Medal.

34. "Edison's Prophecy: A Duplex, Sleepless, Dinnerless World." Edison speculated that earth was receiving wireless signals from outer space. Marshall, "Interview with Thomas Edison," 6. See also Israel, *Edison*, 440–447, 460–462.

35. Hoover's speech was transcribed: "Stenographic Report of the Proceedings of Light's Golden Jubilee."

36. "Stenographic Report," 34.

37. Two views of Menlo Park supplied circa 1980 by Rosemary Papp, historical architect, Greenfield Village.

38. *Detroit Free Press*, October 22, 1929.

39. Upward, *A Home for Our Heritage*, 37–49, 104–107.

40. "Stenographic Report," 34.

41. Israel, *Edison*, 260–362.

42. Lowenthal, *Literature, Popular Culture, and Society*, 110–112.

43. Labor historian Stanley Aronowitz argued that in the new mass culture workers were losing their collective history and "the traditional forms of proletarian culture and everyday life—which gave working class communities their coherence." Aronowitz, *False Promises*, 93.

44. Horkheimer quoted in Jay, *The Dialectical Imagination*, 214.

45. Edison Laboratory Archives, filed under Mina Edison, D1929.

46. Cited in Jay, *The Dialectical Imagination*, 213.

47. Habermas, *Theory of Communicative Action*, 367.

48. "Stenographic Report," 1.

49. "Stenographic Report," 5.

50. "Stenographic Report," 5.

51. "Stenographic Report," 6.

52. "Stenographic Report," 18

53. "Stenographic Report," 30.

54. Einstein's remarks in German were not recorded in the stenographic report, but can be found in a limited-edition publication, *Light's Golden Jubilee*, preserved at the Henry Ford.

55. "To Thomas Alva Edison, On the Occasion of the Dinner at Greenfield, Michigan, October 21, 1929, Celebrating the Golden Jubilee of the Invention of His Incandescent Electric Light." Greenfield, Michigan, 1929, 32–38. Bentley Historical Library, University of Michigan.

56. "Stenographic Report," 27.

57. Edison Laboratory Archives, West Orange, filed under Edison, speeches.

58. Conot, *A Streak of Luck*, 444–445.

59. Barstow, "To Thomas Alva Edison"; "Stenographic Report," 15.

60. Doob, *Propaganda*, 195.

61. Herbert Hoover, "Statement on a National Tribute to Thomas Alva Edison," The American Presidency Project, October 30, 1931, https://www.presidency.ucsb.edu/documents/statement-national-tribute-thomas-alva-edison.

62. Gary Richardson, Alejandro Komai, Michael Gou, and Daniel Park, "Stock Market Crash of 1929," Federal Reserve History, https://www.federalreservehistory.org/essays/stock-market-crash-of-1929.

CHAPTER 10

1. Nye, *American Illuminations*, 70–71, 116–131.

2. See Ball State University Digital Media Repository, https://dmr.bsu.edu/digital/collection/WMRead.

3. Stuckey, *Slave Culture*, 10–11; Floyd, "Ring Shout!"

4. Jaffe, *Joseph Stella's Symbolism*, comments on plate 3.

5. Kasson, "Seeing Coney Island," 97–98.

6. Stella used a similar technique in *Coney Island* (1914), *Spring (The Procession)* (1914–1916), and *Futurist Composition* (1914).

7. Fitzgerald, *Tales of the Jazz Age*, 260–261.

8. Dos Passos, *U.S.A.*, v–vi.

9. Dos Passos, *The Big Money*, 561.

10. Nanney, "John Dos Passos and the Modern(ist) Machine," 9–10. See also her *John Dos Passos and Cinema*.

11. Cited in Wilson, Pilgrim, and Tashjian, *The Machine Age in America*, 30.

12. Tsujimoto, *Images of America*, 33, 107–160.

13. Wilson et al., *The Machine Age in America*, 79–80.

14. Schleier, *The Skyscraper in American Art*, 111–118.

15. Hufford, *Reframing Technology*, 28, 60–61, 89.

16. See French, *When They Hid the Fire*, 122–141.

17. Schatzberg, *Technology*, 72.

18. Schatzberg, *Technology*, 111–112.

19. Schatzberg, *Technology*, 146–151; Mumford, *Technics and Civilization*.

20. Chase, *Men and Machines*, 73, 82.

21. Hufford, *Reframing Technology*, 61–70.

22. Mumford, *Technics and Civilization*, 183.

23. Mumford, *Technics and Civilization*, 343.

24. Mumford, *Technics and Civilization*, 350.

25. Nye, *America as Second Creation*, 1, 10, 105–115, 40, 290–292.

26. Inge, "Is Our Race Disintegrating?," 153.

27. Jordan, *Machine-Age Ideology*, 217.

28. Dell, "An Anti-Feminist Utopia," x.

29. *Ninth Annual Report of the Secretary of Labor*, 125.

30. Sirovich, *Congressional Record*, 9937–9938.

31. Filene, *The Way Out*, 185

32. Filene, *The Way Out*, 222–227.

33. Popp, "Machine Age Communication," 467.

34. Popp, "Machine Age Communication," 470.

35. Siegfried, *America Comes of Age*, 180, 182, 196. See also Duhamel, *America the Menace*.

36. Stead, *The Americanization of the World*.

37. Huxley, "The Outlook for American Culture," 266–267.

38. Lynd and Lynd, *Middletown*, 8–9.

39. Lynd and Lynd, *Middletown*, 40 note.

40. Lynd and Lynd, *Middletown*, 39–52.

41. Nye, *Electrifying America*, 10–11.

42. Lynd and Lynd, *Middletown*, 253

43. Lynd and Lynd, *Middletown*, 228.

44. Lynd and Lynd, *Middletown*, 244–248.

45. Cowan, *More Work for Mother*.

46. Lassiter, *The Other Side of Middletown*.

47. See Google Books Ngram Viewer, "kaleidoscopic,big business," https://books.google.com/ngrams/graph?content=kaleidoscopic%2Cbig+business.

48. Nye, *America as Second Creation*, 283–302.

49. Isenstadt, *Electric Light*, 191; Nye, *American Technological Sublime*, 182–198.

50. Wattenberg, *The Statistical History of the United States*, 716.

51. Nye, *Electrifying America*, 348.

52. Nye, *Consuming Power*, 198.

EPILOGUE

1. Bryan, "The Cross of Gold," 234.

2. See Maria Aspan, "Trump Has a Pro-Cryptocurrency Platform—While Harris Chases the Crypto Vote," NPR, September 25, 2024, https://www.npr.org/2024/09/25/nx-s1-5056183/trump-has-a-pro-cryptocurrency-platform-while-harris-chases-the-crypto-vote.

3. See Donald J. Trump, "Remarks at a Tribal, State, and Local Energy Roundtable Discussion and an Exchange with Reporters," The American Presidency Project, June 28, 2017, https://www.presidency.ucsb.edu/documents/remarks-tribal-state-and-local-energy-roundtable-discussion-and-exchange-with-reporters.

4. See Whitehouse.gov, "President Trump Vows to Usher in Golden Era of American Energy Dominance," June 30, 2017, https://trumpwhitehouse.archives.gov/articles/president-trump-vows-usher-golden-era-american-energy-dominance/.

5. ABC News, "Trump Says 'Drill, Baby, Drill,' but the Record for US Oil Production Isn't His," July 19, 2024, https://abcnews.go.com/Politics/drill-baby-drill-donald-trump-oil-gas-rnc/story?id=112108980.

6. Emmett and Nye, *The Environmental Humanities*, 117–131, 174–176.

7. European Environment Agency, "Share of Energy from Renewable Sources, by Country," March 27, 2024, https://www.eea.europa.eu/en/analysis/maps-and-charts/countries-breakdown-actual-res-progress-15.

8. Nadja Popovich, "How Does Your State Make Electricity, *New York Times*, August 2, 2024, https://www.nytimes.com/interactive/2024/08/02/climate/electricity-generation-us-states.html.

9. "Fossil Fuels Fall Below 50% of US Electricity for the First Month on Record," Ember, April 4, 2025, https://ember-energy.org/latest-updates/fossil-fuels-fall-below-50-of-us-electricity-for-the-first-month-on-record/.

10. "Energy Consumption by Country 2025," World Population Review, accessed September 5, 2025, https://worldpopulationreview.com/country-rankings/energy-consumption-by-country.

BIBLIOGRAPHY

1. Schutz and Luckmann, *The Structures of the Lifeworld*.

2. Bijker, Hughes, and Pinch, *The Social Construction of Technological Systems*; Bijker and Law, *Shaping Technology/Building Society*; Nye, *Technology Matters*, 61–66, 105–107, 185–207; Nye, *Electrifying America*, ix–xi; Nye, *Consuming Power*, 1–6, 94–100; Nye, *America's Assembly Line*, 307.

3. Hughes, *Networks of Power*; Hughes, "Technological Momentum."

4. Nye, "A Model for Heterogeneous Energy Transitions."

5. Smith and Marx, eds., *Does Technology Drive History?*; Nye, *Consuming Power*, 133–139, 150–152.

6. Beard and Beard, *The Rise of American Civilization*.

7. Hoftstadter, *The Age of Reform*.

8. Wiebe, *The Search for Order, 1877–1920*.

9. Chandler, *The Visible Hand*.

10. Brinkley, "Richard Hofstadter's *The Age of Reform*," 472.

11. Schneirov, "Thoughts on Periodizing the Gilded Age," 189–190.

12. Higham, *Hanging Together*, 123.

13. Fuchs, *The American Kaleidoscope*, 276.

14. For critiques of Fuchs, see Gerstle, "Liberty, Coercion, and the Making of Americans."

BIBLIOGRAPHY

BIBLIOGRAPHIC NOTE: This book is partly inspired by the concept of the lifeworld, as developed by Edmund Husserl, by Jürgen Habermas, and by Alfred Schutz and Thomas Luckmann in *The Structures of the Lifeworld*.[1] However, none of these authors focused on the role of technologies in the formation and disruption of lifeworlds. Like many historians of technologies, I consider inventions and systems of machines not to be deterministic but rather social constructions, as I argued in *Electrifying America*, *Consuming Power*, *Technology Matters*, *America's Assembly Line*, and other works.[2] Technological systems may seem deterministic, but they are better understood as culturally shaped constructions that can gradually achieve momentum.[3] In the specific case of energy transitions, inventions are used to disrupt one lifeworld and make possible the construction of a new one. Such transitions develop according to a characteristic structure, but they are so shaped by geographic, political, economic, and social factors that the results vary from one culture to another. The outcomes are not uniform but heterogeneous.[4] Other historians of technology hold similar views, as witnessed by the works cited here by Thomas Hughes, Leo Marx, Eric Schatzberg, Julie Cohn, Christopher Jones, Sandy Isenstadt, Merritt Roe Smith, Richard Hirsh, Martin Melosi, Joel Tarr, Jennifer L. Lieberman, Jonathan Coopersmith, Rebecca Wright, and many others.

Few historians of the Progressive Era seem aware of this scholarship, and their publications seldom pay much attention to technology. Those that do tend to rely on discredited, deterministic ideas such as "Taylorism" and "Fordism."[5] They tend to focus too much on individual entrepreneurs and too little on the social construction of technical change. Early works, notably by Charles and Mary Beard, thought in terms of clashing economic interests.[6] Richard Hofstadter disagreed in *The Age of Reform*, contending that non-economic factors played an important role, notably an agrarian myth of the past and the status anxieties of elites threatened by the new industrial order.[7] Robert Wiebe in *The Search for Order, 1877–1920* reinterpreted the era in terms of modernization theory, in which consolidation of the United States into a single market prompted protest movements, weakened regionalism, stimulated realignments in political organization, and underpinned the rise of cities to dominance.[8] Somewhat in harmony with this approach, business historians, notably Alfred D. Chandler, emphasized how the visible hand of management consolidated American capitalism.[9]

By 1985 historians were moving away from both Hofstadter and his critics. As Alan Brinkley summarized, "Instead of identifying a single, dominant Progressive constituency or a clear, common Progressive program," scholars had begun "to argue for a more pluralistic view that leaves room for many different groups and many different impulses. Some have gone so far as to challenge the existence of a Progressive movement at all; others have attempted to divide progressivism into two distinct impulses; still others have begun to look beyond the particular issues that dominated Progressive rhetoric and to place the phenomenon in the context of a much larger transformation of American political life."[10] The latter included Conn's *The Divided Mind: Ideology and Imagination in America, 1898–1917* and Trachtenberg's *The Incorporation of America*.

Yet not synthesis but specialization predominated. In 2006 Richard Schneirov recalled that by the 1980s fragmentation had "generated an appeal for 'synthesis'" as Thomas Bender called for new narratives "that would transcend 'recent scholarship with its intensive specialization, fragmentation, and preoccupation with groups.' Yet, since then, occasional attempts to synthesize have been stillborn, and for the Gilded Age as well as for the Progressive Era the attempts at synthesis seem to have reached a cul-de-sac."[11] Jamie L. Pietruska noted, "A single long Progressive Era emerged in Michael McGerr's *A Fierce Discontent: The Rise and Fall of the Progressive Movement in America, 1870–1920* and in Nell Irvin Painter's *Standing at Armageddon: A Grassroots History of the Progressive Era*." More recently, Jackson Lears synthesized previous scholarship in *Rebirth of a Nation*. But these works contained little technological history and did not discuss the great energy transition.

Two historians adopted the kaleidoscopic metaphor. John Higham suggested it in *Hanging Together: Unity and Diversity in American Culture*.[12] Lawrence Fuchs used it in *The American Kaleidoscope: Race, Ethnicity, and the*

Civic Culture. He argued that this metaphor "best explains the dynamic of ethnicity."[13] He saw an increasing acceptance of all races, creeds, and ethnic groups to full participation in democracy. Fuchs built on Tocqueville's analysis of the United States as a "culture of politics" rather than one based on ancestry, race, or religion. Like Jane Addams, Randolph Bourne, and John Dewey, Fuchs rejected the melting pot metaphor in favor of a "voluntary pluralism" in which all groups participate as equals in public life, while remaining free to differ in religion and in private life. Fuchs anticipated that Asian immigrants, Hispanics, African Americans, and Native Americans would join this "voluntary pluralism" in a harmonious multicultural society. Many rejected this optimistic argument, however, including Irving Howe, Lisabeth Cohen, Gwendolyn Mink, Roy Rosensweig, Gary Gerstle, David R. Roediger, and Michael Rogin.[14] They did not agree that Americanization was voluntary, liberating, or facilitated the construction of new identity. Rather, they argued that white consensus was to a considerable degree based upon exclusion, and that the pressures of class, race, and gender constrained possible identities in a hegemonic culture. I agree with this critique, and in chapter 5 I place it within the framework of the great energy transition.

Adams, Henry. *The Degradation of the Democratic Dogma*. London: Macmillan, 1919.

Adams, Henry. *The Education of Henry Adams*. New York: Modern Library, 1931.

Adams, James Truslow. *Our Business Civilization*. New York: Albert & Charles Boni, 1929.

Adas, Michael. *Machines as the Measure of Men*. Ithaca, NY: Cornell University Press, 1989.

Aldrich, Thomas Bailey. "Unguarded Gates," *Atlantic Monthly*, July 1892.

"Amusements for Park Resorts." *Street Railway Journal* 25, no. 10 (1905).

Anderson, Nancy K. "The Kiss of Enterprise: The Western Landscape as Symbol and Resource." In *The West as America*, ed. W. Treuttner, 237–284. Washington, DC: Smithsonian Institution Press, 1991.

Aron, Cindy S. *Working at Play: A History of Vacations in the United States*. New York: Oxford University Press, 1999.

Aronowitz, Stanley. *False Promises*. New York: McGraw-Hill, 1973.

Babcock, Louis. *Our American Resorts*. Washington, DC: National News Bureau, 1884.

Bain, David Haward. *Empire Express: Building the First Transcontinental Railroad*. New York: Penguin Books, 1999.

Baldwin, Peter. "When Cities Treated Cars as Dangerous Intruders." The MIT Press Reader, July 25, 2023. https://thereader.mitpress.mit.edu/when-cities-treated-cars-as-dangerous-intruders/.

Barringer, Felicity. "Despite Cleanup at Mine, Dust and Fear Linger," *New York Times*, April 12, 2004.

Barstow, Walter S. "To Thomas Alva Edison," "Stenographic Report of the Proceedings of Light's Golden Jubilee," Greenfield Village, The Henry Ford.

Bates, Charles A. *The Art and Literature of Business*. New York: Bates Publishing, 1902.

Bazerman, Charles. *The Languages of Edison's Light*. Cambridge, MA: MIT Press, 1999.

Beard, Charles, and Mary Beard. *The Rise of American Civilization*. New York: Macmillan, 1927.

Beard, George. *American Nervousness*. New York. G. P. Putnam's Sons, 1880.

Bellamy, Edward. *Looking Backward*. Boston: Houghton Mifflin, 1889.

Beniger, James R. *The Control Revolution: Technological and Economic Origins of the Information Society*. Cambridge, MA: Harvard University Press, 1986.

Benjamin, Walter. "The Work of Art in the Age of Mechanical Reproduction." In *Illuminations: Essays and Reflections*, ed. Hannah Arendt, trans. Harry Zohn, 217–251. New York: Schocken Books, 1969.

Berry, Wendell. *The Unsettling of America: Culture & Agriculture*. San Francisco: Sierra Club Books, 1986.

Bierce, Ambrose. "Ashes of the Beacon." In *The Collected Works of Ambrose Bierce*, vol. 1, 21–22. https://books.apple.com/dk/book/the-collected-works-of-ambrose-bierce-volume-1/id507936000.

Bijker, Wiebe E., Thomas P. Hughes, and Trevor Pinch, eds. *The Social Construction of Technological Systems: New Directions in the Sociology and History of Technology*. Cambridge, MA: MIT Press, 1987.

Bijker, Wiebe E., and John Law, eds. *Shaping Technology/Building Society: Studies in Sociotechnical Change*. Cambridge, MA: MIT Press, 1992.

Bingham, Charles S. "Bibliography of American Editions of Robinson Crusoe to 1830." *Proceedings of the American Antiquarian Society*, 67, no. 2 (October 1957): 137–183.

Bix, Amy. *Inventing Ourselves Out of Jobs? America's Debate over Technological Unemployment*. Baltimore: Johns Hopkins University Press, 2000.

Blaszczyk, Regina Lee. *Imagining Consumers: Design and Innovation from Wedgwood to Corning*. Baltimore: Johns Hopkins University Press, 2000.

Blonheim, Menahem. *News over the Wires: The Telegraph and the Flow of Public Information in America, 1844–1897*. Cambridge, MA: Harvard University Press, 1994.

Blumberg, Barbara. "A National Monument Emerges." In New York Public Library, *Liberty: The French Statue in Art and History*. New York. Harper & Row, 1986.

Bohlen, Francis H. "Assumption of Risk." *Harvard Law Review* 20, no. 1 (November 1906): 14–34.

Boime, Albert. *The Magisterial Gaze: Manifest Destiny and American Landscape Painting*. Washington, DC: Smithsonian Institution Press, 1991.

Bok, Edward William. *The Americanization of Edward Bok*. New York: Scribner's, 1920.

Boorstin, Daniel. *The Americans: The Democratic Experience*. New York: Vintage Books, 1973.

Bourne, Randolph. "The Social Order in an American Town," *The Atlantic*, February 1913, 227–236.

Bowen, John Wesley Edward. "An Appeal to the King: The Address Delivered on Negro Day in the Atlanta Exposition, October 21, 1895." In *The New Negro: Readings on Race, Representation, and African American Culture, 1892–1938*, ed. Henry Louis Gates Jr., and Gene Andrew Jarrett. Princeton, NJ: Princeton University Press, 2007.

Brewster, David. *A Treatise on the Kaleidoscope*. Edinburgh: Archibald Constable & Co., 1819.

Brinkley, Alan. "Richard Hofstadter's *The Age of Reform*: A Reconsideration." *Reviews in American History* 13, no. 3 (September 1985): 462–480.

Bryan, William Jennings. "The Cross of Gold." In *Official Proceedings of the Democratic National Convention Held in Chicago, Illinois, July 1896*. Logansport, IN: The Democratic Party, 1896.

Bryce, James. *The American Commonwealth*. 2 vols. New York: Macmillan Company, 1919.

"Buffalo Bill's Review of the Representative Rough Riders of the World." *Emporia Daily Gazette*, September 13, 1898.

Burch, Henry Reed, and S. Howard Patterson. *Problems of American Democracy: Political, Economic, Social*. New York: Macmillan, 1922.

Cameron, William E. *The World's Fair: Being a Pictorial History of the Columbian Exposition*. Chicago: P. D. Farrell, 1893.

Carnegie, Andrew. "The Gospel of Wealth." *North American Review*, June 1889, 653–664.

Cather, Willa. "Amusements." *Nebraska State Journal*, September 28, 1894, 6.

Cawelti, John. *Apostles of the Self-Made Man*. Chicago: University of Chicago Press, 1965.

"Celebration of the Completion of the Erie Canal." *Albany Daily Advertiser*, October 11, 1825.

Census Bureau. *Historical Statistics of the United States, Colonial Times to 1970*. Washington, DC: US Government Printing Office, 1976.

"Centennial Exhibition." *Baltimore Sun*, May 24, 1876.

"Centennial Letter." *New Hampshire Sentinel*, May 18, 1876.

Chandler, Alfred. *The Visible Hand*. Cambridge, MA: Harvard University Press, 1977.

Chase, Stuart. *Men and Machines*, New York: Macmillan, 1929.

Chauncey, George. *Gay New York: Gender, Urban Culture and the Making of the Gay Male World, 1890–1940*. New York: Basic Books, 1994.

"The City of Pittsburgh." *Harper's New Monthly Magazine* 62 (1880–1881):49–68.

Cohn, Julie. *The Grid: Biography of an American Technology*. Cambridge, MA: MIT Press, 2017.

Conn, Peter. *The Divided Mind: Ideology and Imagination in America, 1898–1917*. Cambridge: Cambridge University Press, 1988.

Conot, Robert. *A Streak of Luck: The Life and Legend of Thomas Alva Edison*. New York: Seaview Books, 1979.

Conwell, Russell H. *Acres of Diamonds*. New York: Harper Brothers, 1915.

Cowan, Ruth Schwartz. *More Work for Mother*. New York: Basic Books, 1983.

Cox, James. "Regionalism: A Diminished Thing." In *Columbia Literary History of the United States*, ed. Emory Elliott. New York: Columbia University Press, 1988.

Coyne, Heather Campbell, and Joyce K. Schiller. "John Sloan's Urban Encounters." In *John Sloan's New York*. Wilmington: Delaware Art Museum, 2007.

Crabb, A. Richard. *The Hybrid-Corn Makers: Prophets of Plenty*. New Brunswick, NJ: Rutgers University Press, 1947.

Cram, Ralph Adams. *Walled Towns*. Boston: Marshall Jones, 1919.

Cronon, William. *Nature's Metropolis: Chicago and the Great West*. New York: Norton, 1991.

Cullis, Philip. "The Limits of Progressivism: Louis Brandeis, Democracy and the Corporation." *Journal of American Studies* 30, no. 3 (1996): 381–404.

"Current Magazines." *New York Times*, May 30, 1926, B19.

Czitrom, Daniel. J. *Media and the American Mind: From Morse to McLuhan*. Chapel Hill: University of North Carolina Press, 1982.

Danbom, David B. *Born in the Country: A History of Rural America*. Baltimore: Johns Hopkins University Press, 2006.

Daniels, Roger. *Asian America: Chinese and Japanese in the United States since 1850*. Seattle: University of Washington Press, 1988.

Dayan, Daniel. "Beyond Media Events: Disenchantment, Derailment, Disruption." In *Owning the Olympics: Narratives of the New China*, ed. Monroe Price and Daniel Dayan, 391–402. Ann Arbor: University of Michigan Press, 2008.

Dayan, Daniel, and Elihu Katz. *Media Events: The Live Broadcasting of History*. Cambridge, MA: Harvard University Press, 1992.

"Dedicated! The Act Which Completes the Washington Monument," *Washington Post*, February 22, 1885, 1–3.

Defoe, Daniel. *An Essay upon Projects*. London: Printed by R. R. for Tho. Cockerill, 1697.

Defoe, Daniel. *The Further Adventures of Robinson Crusoe* (1719). London: Pickering & Chatto, 2008–2009.

Defoe, Daniel. *A General History of Discoveries and Improvements* (1725–1726) and *An Essay upon Literature* (1726). Ed. P. N. Furbank. London: Pickering & Chatto, 2001–2002.

Defoe, Daniel. *Robinson Crusoe*. New York: Barnes and Noble, 2003.

Dell, Floyd. "An Anti-Feminist Utopia," *Current History*, November 1927, x–xiv.

Derickson, Alan. "'On the Dump Heap': Employee Medical Screening in the Tri-State Zinc-Lead Industry, 1924–1932," *Business History Review* 82 (Winter 1988): 656–677.

Dewey, John. "Americanism and Localism." *The Dial* 68 (1920): 684–688.

"Display of Fireworks at the Columbian Exposition." *Scientific American* 69, no. 23 (December 2, 1893): 359.

Donovan, Bryan L. "Framing and Strategy: Explaining Differential Longevity in the Woman's Christian Temperance Union and the Anti-Saloon League." *Sociological Inquiry* 65, no. 2 (April 1995): 143–154.

Doob, Leonard W. *Propaganda: Its Psychology and Technique*. New York: Henry Holt, 1935.

Dos Passos, John. *U.S.A.* New York: Modern Library, 1937.

Douglas, Mary, and Baron Isherwood. *The World of Goods*. New York: 1979.

Douglas, Susan. *Inventing American Broadcasting, 1899–1922*. Baltimore: Johns Hopkins University Press, 1987.

Dreiser, Theodore. *A Hoosier Holiday*. New York: Oxford University Press, 1932.

Dreiser, Theodore. *Sister Carrie*. New York: Penguin, 1981.

Du Bois, W. E. B. *The Souls of Black Folk*. Chicago: McClurg, 1903.

Duhamel, Georges. *America the Menace: Scenes from the Life of the Future*. Boston: Houghton Mifflin, 1931.

Dyer, Frank L., and Thomas C. Martin, with William H. Meadowcroft. *Edison: His Life and Inventions*. New York: Harper & Bros., 1910. Revised edition 1929.

"Edison in His Workshop." *Harper's Weekly* 23 (August 2, 1879): 607.

"Edison's Prophecy: A Duplex, Sleepless, Dinnerless World." *Literary Digest*, November 14, 1914, 966–968.

Edwards, Alma M. *Comparative Occupation Statistics for the United States, 1870–1940*. Sixteenth Census of the United States, Washington, DC: United States Government Printing Office, 1943.

Eliot, T. S. *Ash Wednesday*. London: Faber and Faber, 1930.

Eliot, T. S. *Notes Toward the Definition of Culture*. London: Faber & Faber, 1948.

Eliot, T. S. *The Waste Land*. London: Boni and Liveright, 1922.

Emerson, Ralph Waldo. "Quotation and Originality." In *The Complete Works of Ralph Waldo Emerson: Letters and Social Aims*, vol. 8. Boston: Houghton, Mifflin, 1911.

Emmett, Robert E., and David E. Nye. *The Environmental Humanities, A Critical Introduction*. Cambridge, MA: MIT Press, 2017.

Enck, Henry S. "Tuskegee Institute and Northern White Philanthropy: A Case Study in Fund Raising, 1900–1915." *Journal of Negro History* 65, no. 4 (Fall 1980): 336–348.

Ewen, Stuart. *Captains of Consciousness: Advertising and the Social Roots of the Consumer Culture*. New York: McGraw-Hill, 1975.

Faulkner, William. *Requiem for a Nun*. New York: Random House, 1951.

Ferris, George T., ed. *Our Native Land; or, Glances at American Scenery and Places, with Sketches of Life and Adventure*. New York: D. Appleton, 1882.

Filene, Edward A. *The Way Out*. New York: Doubleday, Page, 1924.

Fischer, Claude S. *American Calling: A Social History of the Telephone*. Berkeley: University of California Press, 1992.

Fiske, Stephen. *Off-Hand Portraits of Prominent New Yorkers*. New York: George Lockwood & Son, 1884.

Fitzgerald, F. Scott. *Tales of the Jazz Age*. New York: Charles Scribner's, 1922.

Fitzgerald, F. Scott, *The Great Gatsby*. New York: Charles Scribner's, 1925.

Floyd, Samuel A., Jr. "Ring Shout! Literary Studies, Historical Studies, and Black Music Inquiry." *Black Music Research Journal* 22 (2002): 49–70.

Føllesdal, Dagfinn. "The *Lebenswelt* in Husserl." In *Language, Knowledge and Intentionality*, ed. Leila Haaparanta et al., 123–143. Helsinki: Acta Philosophica Fennica, 1990.

Foner, Eric. *The Second Founding: How the Civil War and Reconstruction Remade the Constitution*. New York: W. W. Norton, 2019.

Foner, Eric. "Why Is There No Socialism in the United States?" *History Workshop*, no. 17 (Spring 1984): 57–80.

"A Food Creator: Edison Invents a Machine That Will Feed the Human Race." *New York Daily Graphic*, April 1, 1878.

Ford, Henry, and Samuel Crowther. *My Life and Work*. London: William Heinemann, [1922] 1925.

Ford, Lucy. *Female Robinson Crusoe: A Tale of the American Wilderness*. New York: Printed by J. W. Bell, 1837.

Frank, Robin Jaffee. *Coney Island: Visions of an American Dreamland, 1861–2008*. New Haven, CT: Yale University Press, 2015.

Frank, Waldo. *Our America*. New York: Boni and Liveright, 1919.

Frazer, Douglas. *Perseverance Island; or, The Robinson Crusoe of the Nineteenth Century*. Boston: Lee and Shepard, 1885.

French, Daniel. *When They Hid the Fire*. Pittsburgh: University of Pittsburgh Press, 2017.

Freund, Daniel. *American Sunshine: Diseases of Darkness and the Quest for Natural Light*. Chicago: University of Chicago Press, 2012.

Friedel, Robert. *The Culture of Improvement*. Cambridge, MA: MIT Press, 2007.

Frost, Robert. "Preface." In *The Collected Poems of Robert Frost*. New York: Henry Holt, 1939.

Frost, Stanley. "Radio Dreams That Can Come True." *Collier's* 69 (June 10, 1922).

Fuchs, Lawrence H. *The American Kaleidoscope: Race, Ethnicity, and the Civic Culture*. Hanover, NH: Wesleyan University Press, 1990.

Funk, Cary. "Republicans' Views on Evolution." Pew Research Poll, January 3, 2014.

Garcia, Monica Isabel. "A History of Mexican and Mexican Americans in the Coal Mining Regions of the Rocky Mountain West, 1880–1928." Dissertation, University of Santa Barbara, 2012.

Gates, Henry Louis, Jr., and Gene Andrew Jarrett, eds. *The New Negro: Readings on Race, Representation, and African American Culture, 1892–1938*. Princeton, NJ: Princeton University Press, 2007.

Gatewood, John B., and Catherine M. Cameron. "Battlefield Pilgrims at Gettysburg National Military Park." *Ethnology* 43, no. 3 (2004): 193–216.

Gerstle, Gary. "Liberty, Coercion, and the Making of Americans." *Journal of American History* 84, no. 2 (September 1997): 524–558.

Gibson, Campbell. "Population of the 100 Largest Cities and Other Urban Places in the United States: 1790–1990." Working Paper Number POP-WP027. Washington, DC: Census Bureau, 1998.

Gillette, King Camp. *The Human Drift*. Boston: New Era, 1894.

Gingrich, Newt. *To Renew America*. New York: HarperCollins, 1994.

Gleick, Janes. *Faster*. New York: Abacus, 1999.

Goldberg, David J. *Discontented America: The United States in the 1920s*. Baltimore: Johns Hopkins University Press, 1999.

Goldfield, David, and Blaine Brownell. *Urban America: From Downtown to No Town*. Boston: Houghton Mifflin, 1979.

Gordon, Robert. *The Rise and Fall of American Growth*. Princeton, NJ: Princeton University Press, 2016.

"Grand Canal Celebration." *Utica Sentinel*, November 8, 1825.

Grant, Madison. *The Passing of the Great Race*. New York: Scribner's, 1916.

Grapard, Ulla. "Robinson Crusoe: The Quintessential Economic Man?" *Feminist Economics* 1 (1995): 33–52.

Grapard, Ulla, and Gillian Hewitson, eds. *Robinson Crusoe's Economic Man: A Construction and Deconstruction*. London: Routledge, 2011.

Grey, Zane. *Riders of the Purple Sage*. New York: Harper & Brothers, 1912.

Groth, Helen. "Kaleidoscopic Vision and Literary Invention in an 'Age of Things': David Brewster, Don Juan, and 'A Lady's Kaleidoscope.'" *ELH* 74 (2007): 217–237.

Gunn, R. T. "Summer Resorts." *Street Railway Journal* 25, no. 10 (1905): 360.

Gusfield, Joseph. *Symbolic Crusade: Status Politics and the American Temperance Movement*. Urbana: University of Illinois Press, 1963.

Habermas, Jürgen. *Theory of Communicative Action*, vol. 2. Boston: Beacon Press, 1987.

Hales, Peter Bacon. *Silver Cities: The Photography of American Urbanization, 1839–1915*. Philadelphia: Temple University Press, 1984.

Haller, Frank. "Riding in a German Airship." *West Liberty Index* (Iowa), June 22, 1916.

Hargrove, June. "Unveiling the Colossus." In New York Public Library, *Liberty: The French Statue in Art and history*. New York: Harper & Row 1986.

Harper, Ida Husted. *The Life and Work of Susan B. Anthony*. 3 vols. North Stratford, NH: The Ayer Company, 1998.

Hayes, Elinor. "The Employees' Publication." *University Journal of Business* (University of Chicago) 1, no. 1 (November 1922).

Hays, Samuel P. *Conservation and the Gospel of Efficiency*. New York: Atheneum, 1979.

Higham, John. *Hanging Together: Unity and Diversity in American Culture*. New Haven, CT: Yale University Press, 2001.

Hirsch, James S. *Riot and Remembrance: The Tulsa Race War and Its Legacy*. New York. Houghton Mifflin, 2002.

Hirsh, Richard F. *Powering American Farms: The Overlooked Origins of Rural Electrification*. Baltimore: Johns Hopkins University Press, 2022.

Hirschman, Charles, and Elizabeth Mogford. "Immigration and the American Industrial Revolution from 1880 to 1920." *Social Science Research* 38, no. 4 (December 2009): 897–920.

Hobbs, Allyson. *A Chosen Exile: A History of Racial Passing in American Life*. Cambridge, MA: Harvard University Press, 2014.

Hochschild, Adam. *American Midnight: The Great War. A Violent Peace and Democracy's Forgotten Crisis*. New York: Mariner Books, 2022.

Hofstadter, Richard. *The Age of Reform*. New York: Vintage, 1955.

Hofstadter, Richard. "Ulrich B. Phillips and the Plantation Legend." *Journal of Negro History* 29, no. 2 (April 1944): 109–124.

Hornaday, William T. "The Extermination of the American Bison, with a Sketch of Its Discovery and Life History." Part Two, Report of the United States National Museum for the Year Ending June 30, 1887. Washington, DC: Smithsonian Institution, 1887.

Horwitz, Robert Britt. *The Irony of Regulatory Reform: The Deregulation of Telecommunications*. New York: Oxford University Press, 1989.

Hounshell, David A. *From the American System to Mass Production, 1800–1932*. Baltimore: Johns Hopkins University Press, 1984.

Howells, William Dean. *A Traveler from Altruria*. New York: Harper & Brothers, 1894.

Hufford, Kent. *Reframing Technology: An Intellectual History of Thinking Beyond the Machine*. Berlin: De Gruyter, 2024.

Hughes, Thomas P. "Afterword." *Annales historiques de l'électricité* 2 (2004): 173–176.

Hughes, Thomas P. *American Genesis*. New York: Penguin, 1989.

Hughes, Thomas P. *Networks of Power*. Baltimore: Johns Hopkins University Press, 1983.

Hughes, Thomas P. "Technological Momentum." In *Does Technology Drive History?*, ed. Merritt Roe Smith and Leo Marx. Cambridge, MA: MIT Press, 1994.

"The Human Coalburner." *Railway Times*, September 15, 1860, 12, 37.

Hunter, Louis C. *Waterpower in the Century of the Steam Engine*. Charlottesville: University of Virginia Press, 1979.

Huxley, Aldous. "The Outlook for American Culture: Some Reflections in a Machine Age." *Harper's*, August 1927, 265–270.

Hyde, Anne Farrar. *An American Vision: The Far Western Landscape and National Culture, 1820–1920*. New York: New York University Press, 1990.

Ihde, Don. *Technology and the Lifeworld: From Garden to Earth*. Bloomington: Indiana University Press, 1990.

Inge, W. R. "Is Our Race Disintegrating?" *The Living Age* 332 (1927): 153–156.

Ingram, David. *Habermas: Introduction and Analysis*. Ithaca, NY: Cornell University Press, 2011.

Isenstadt, Sandy. *Electric Light: An Architectural History*. Cambridge, MA: MIT Press, 2018.

Israel, Paul. *Edison: A Life of Invention*. New York: John Wiley, 1998.

Jackson, Donald. *Pastoral and Monumental; Dams, Postcards, and the American Landscape*. Pittsburgh: Pittsburgh University Press, 2013.

Jackson, J. B. "The Abstract World of the Hot-Rodder." In *Changing Rural Landscapes*, ed. Ervin H. Zube and Margaret J. Zube, 140–151. Amherst: University of Massachusetts Press, 1977.

Jaffe, Irma B. *Joseph Stella's Symbolism*. San Francisco: Pomegranate Artbooks, 1994.

Jakle, John A. *The Tourist: Travel in Twentieth-Century North America*. Lincoln: University of Nebraska Press, 1985.

James, Henry. *Hawthorne*. London: Macmillan, 1879.

Jameson, Fredric. "Reification and Utopia in Mass Culture." *Social Text*, Winter 1979, 130–149.

Jay, Martin. *The Dialectical Imagination: A History of the Frankfurt School and the Institute for Social Research, 1923–1950*. London: Heinemann, 1976.

Jehl, Francis. *Menlo Park Reminiscences*, 3 vols. Dearborn, MI: Edison Institute, 1937.

Jerome Redding Company, *Catalog*, 1875.

Johnson, James Weldon. "The Riots." *Crisis Magazine*, September 1919, 241–243.

Jones, Christopher F. *Routes of Power*. Cambridge, MA: Harvard University Press, 2014.

Jordan, John M. *Machine-Age Ideology: Social Engineering and American Liberalism, 1911–1939*. Chapel Hill: University of North Carolina, 1994.

Josephson, Matthew. *Edison: A Biography*. New York: McGraw Hill, 1959.

Kammen, Michael. *In the Past Lane: Historical Perspectives on American Culture*. New York: Oxford University Press, 1997.

Kaplan, Amy. "Nation, Region, Empire." In *The Columbia History of the American Novel*, ed. Emory Elliott. New York: Columbia University Press, 1991.

Kasson, John F. *Amusing the Million: Coney Island at the Turn of the Century*. New York: Hill and Wang, 1978.

Kasson, John F. "Seeing Coney Island, Seeing Culture: Joseph Stella's Battle of Lights," *Yale Journal of Criticism* 11, no. 1 (1998): 95–101.

Kazin, Michael. *The Populist Persuasion: An American History*. New York: Basic Books, 1995.

Kemble, Fanny. *Further Records, 1848–1883: A Series of Letters*. London: R. Bentley and Son, 1890.

Kern, Stephen. *The Culture of Space and Time, 1880–1918*. Cambridge, MA: Harvard University Press, 1983.

Kline, Ronald R. *Consumers in the Country: Technology and Social Change in Rural America*. Baltimore: Johns Hopkins University Press, 2000.

Kolko, Gabriel. *The Triumph of Conservatism*. Chicago: Quadrangle Books, 1967.

Kouwenhoven, John A. *Half a Truth Is Better than None*. Chicago: University of Chicago Press, 1982.

Kracauer, Siegfried. *The Mass Ornament*. Cambridge: MA: Harvard University Press, 1995.

Krugman, Paul. "Increasing Returns and Economic Geography." *Journal of Political Economy* 99, no. 3. (1991): 483–499.

Krugman, Paul. "Superstar Cities." *New York Times*, September 1, 2023.

Kuta, Sara. "The Man Who Invented Water Skiing." *Smithsonian Magazine*, July 1, 2022.

Kyle, Charles Wesley. *Yosemite, the World's Wonderland*. San Francisco: Chase & Rae, 1915.

Lassiter, Luke E., ed. *The Other Side of Middletown: Exploring Muncie's African American Community*. Walnut Creek, CA: AltaMira Press, 2004.

Layton, Edwin T., Jr. *The Revolt of the Engineers: Social Responsibility and the American Engineering Profession*. Cleveland: Case Western Reserve University Press, 1971.

Leach, William. *Land of Desire: Merchants, Power, and the Rise of a New American Culture*. New York: Pantheon, 1993.

Lears, Jackson. *Fables of Abundance: A Cultural History of Advertising in America*. New York: Basic Books, 1994.

Lears, Jackson. *No Place of Grace: Anti-Modernism and the Transformation of American Culture, 1880–1920*. New York: Pantheon, 1981.

Lears, Jackson. *Rebirth of a Nation: The Making of Modern America*. New York: HarperCollins, 2009.

LeCaine, Timothy J. *Mass Destruction: The Men and Giant Mines That Wired America and Scarred the Planet*. New Brunswick, NJ: Rutgers University Press, 2009.

Leccese, Stephen. "John D. Rockefeller, Standard Oil, and the Rise of Corporate Public Relations in Progressive America, 1902–1908." *Journal of the Gilded Age and Progressive Era* 16, no. 3 (2017): 245–263.

Leiserson, William M. *Adjusting Immigrant and Industry*. New York: Harper & Brothers, 1924.

Lerman, Nina. "Industrial Genders, Constructing Boundaries." In *Gender and Technology: A Reader*, ed. Nina Lerman, Ruth Oldenziel, and Arwen P. Mohun, 123–143. Baltimore: Johns Hopkins University Press, 2003.

Levine, Philippa. *Eugenics: A Very Short Introduction*. Oxford: Oxford University Press, 2017.

Lieberman, Jennifer L. *Power Lines: Electricity in American Life and Letters, 1882–1952*. Cambridge, MA: MIT Press, 2017.

Linenthal, Edward Tabor. *Sacred Ground: Americans and Their Battlefields*. Urbana: University of Illinois Press, 1993.

Lipow, Arthur. *Authoritarian Socialism in America: Edward Bellamy and the Nationalist Movement*. Berkeley: University of California Press, 1982.

Lorimer, George Horace. "Courage and Common Sense." *Saturday Evening Post* 191, no. 33 (February 15, 1919): 20.

Lowenthal, Leo. *Literature, Popular Culture, and Society*. Englewood Cliffs, NJ: Prentice Hall, 1961.

Luckiesh, Matthew. *Artificial Light: Its Influence on Civilization*. New York: Century, 1920.

Lueck, Dean. "The Extermination and Conservation of the American Bison." *Journal of Legal Studies* 31, no. S2 (2002): S609–S652.

Lynd, Robert S., and Helen Merrell Lynd. *Middletown: A Study in Modern American Culture*. New York: Harcourt, Brace and World, 1956.

Magill, Edward H. "The Close of the Exposition." *Friends' Intelligencer and Journal*, November 30, 1889, 757.

Mann, Donald. "Telegraphing of Election Returns." *American Telegraph Magazine*, November 1, 1852. 74–78.

Marchand, Roland. *Advertising the American Dream: Making Way for Modernity, 1920–1940*. Berkeley: University of California Press, 1985.

Marchand, Roland. *Creating the Corporate Soul: The Rise of Public Relations and Corporate Imagery in American Big Business*. Berkeley: University of California Press, 1998.

Markowitz, Gerald, and David Rosner. "'The Street of Walking Death': Silicosis, Health, and Labor in the Tri-State Region, 1900–1950." *Journal of American History* 77, no. 2 (September 1990): 525–552.

Marsh, George Perkins. *Man and Nature*. Seattle: University of Washington Press, 2003.

Marshall, Edward. "Interview with Thomas Edison." *Springfield Union*, January 25, 1920, 6.

Marshall, Logan. *Seeing America*. Philadelphia: John C. Winston, 1915.

Martin, Alma Borth. *A Vocabulary Study of "The Gilded Age."* Introduction by Robert L. Ramsay, foreword by Hamlin Garland. Webster Groves, MO: Mark Twain Society, 1930.

Martin, Edward Winslow. *The Secrets of the Great City*. Philadelphia: Jones Brothers, 1868.

Martin, Ronald. *American Literature and the Universe of Force*. Durham, NC: Duke University Press, 1981.

Marx, Karl. *Capital: A Critical Analysis of Capitalist Production*, vol 1. London: Swan Sonnenschein, Lowrey, 1887.

Marx, Leo. *The Machine in the Garden*. New York: Oxford University Press, 1965.

Marx, Leo. "Technology: The Emergence of a Hazardous Concept." *Technology and Culture* 51, no. 3 (July 2010): 561–577.

McCabe, James. *New York by Gaslight*. New York: Arlington House, [1882] 1984.

McClelland, Linda Flint. *Building the National Parks, Historic Landscape Design and Construction*. Baltimore: Johns Hopkins University Press, 1998.

McClure, J. B. *Edison and His Inventions*. Chicago: Rhodes and McClure, 1879.

McCracken, Grant. *Culture and Consumption*. Bloomington: University of Indiana Press, 1990.

McCraw, Thomas K. *Prophet of Innovation: Joseph Schumpeter and Creative Destruction*. Cambridge, MA: Harvard Belknap Press, 2007.

McCrossen, Alexis. *Marking Modern Times: A History of Clocks, Watches, and Other Timekeepers in American Life*. Chicago: University of Chicago Press, 2013.

McCullough, Robert L. *Old Wheelways: Traces of Bicycle History on the Land*. Cambridge, MA: MIT Press, 2015.

McKendrick, Neil, J. H. Plumb, and John Brewer. *The Birth of a Consumer Society: The Commercialization of Eighteenth Century England*. Bloomington: Indiana University Press, 1982.

Meikle, Jeffrey L. *Twentieth Century Limited: Industrial Design in America, 1925–1939*. Philadelphia: Temple University Press, 1979.

Meinig, D. W. *The Shaping of America: Global America, 1915–2000*. New Haven, CT: Yale University Press, 2004.

Melosi, Martin V. *Effluent America: Cities, Industry, Energy, and the Environment*. Pittsburgh: University of Pittsburgh Press, 2001.

Mendenhall, John. *American Trademarks, 1930–1950*. New York: Art Direction Book Company, 1983.

Meyer, Donald. *The Positive Thinkers: Religion as Pop Psychology*. New York: Pantheon, 1980.

Meyer, Jeffrey F. *Myths in Stone: Religious Dimensions of Washington D.C.* Berkeley: University of California Press, 2001.

Miller, George. *Picture Postcards in the United States, 1893–1918*. New York: Crown, 1976.

Missal, Alexander. *Seaway to the Future: American Social Visions and the Construction of the Panama Canal*. Madison: University of Wisconsin Press, 2008.

Mitchell, W. T. J. *Iconology: Image, Text, Ideology*. Chicago: University of Chicago Press, 1986.

Mock, C. J. "Rainfall in the Garden of the United States Great Plains, 1870–1889." *Climatic Change* 44 (2000): 173–195.

Montgomery, David. *Citizen Worker*. Cambridge: Cambridge University Press, 1993.

Montgomery, David. *Workers' Control in America*. Cambridge: Cambridge University Press, 1979.

Morris, Richard B. *Encyclopedia of American History*. New York: Harper & Row, 1970.

Mumford, Lewis. *Technics and Civilization*. New York: Harcourt, Brace, 1934.

Mutel, Cornelia F. *The Enchanted Horizon: The History of Nature in Iowa*. Iowa City: University of Iowa Press, 2008.

Nail, Thomas. *Marx in Motion*. Oxford: Oxford University Press, 2020.

Nanney, Lisa. *John Dos Passos and Cinema*. Liverpool: Liverpool University Press, 2019.

Nanney, Lisa. "John Dos Passos and the Modern(ist) Machine." *American Literatures* 2, no. 2 (2022): 9–31.

Nasaw, David. *Going Out: The Rise and Fall of Public Amusements*. New York: Basic Books, 1993.

Nash, Roderick. *The Nervous Generation: American Thought, 1917–1930*. Chicago: Rand McNally, 1970.

Nash, Steve. *Grand Canyon for Sale*. Berkeley: University of California Press, 2017.

National Association for the Advancement of Colored People. *Thirty Years of Lynching in the United States, 1889–1918*. New York: National Association for the Advancement of Colored People, 1919.

National Park Service, "Many Colored Canyon." In *Rules and Regulations, Yellowstone National Park*. Washington, DC: Government Printing Office, 1920.

National Research Council. *Assessing and Managing the Ecological Impacts of Paved Roads*. Washington, DC: National Academies Press, 2005.

Nevins, Alan. *John D. Rockefeller: The Heroic Age of American Enterprise*. New York: Scribner's, 1940.

New York Illustrated: A Pictorial Delineation of Street Scenes, Buildings, River Views, and Other Features of the Great Metropolis. New York: Appleton, 1881.

Newell, Frederick Haynes. *Irrigation in the United States*. New York: Thomas Y. Crowell, 1902.

"Niagara Falls from a New Point of View." *Scientific American* 105, no. 11 (September 9, 1911): 227.

Nichols, Tom. "Why Schumpeter Was Right: Innovation, Market Power, and Creative Destruction in 1920s America." *Journal of Economic History*, 63, no. 4 (December 2003): 1023–1058.

Ninth Annual Report of the Secretary of Labor. Washington, DC: Government Printing Office, 1921.

Nixon, Richard. "Special Message to the Congress Proposing Patent Modernization and Reform Legislation." September 27, 1973. https://www.presidency.ucsb.edu/documents/special-message-the-congress-proposing-patent-modernization-and-reform-legislation.

Nixon, Rob. *Slow Violence and the Environmentalism of the Poor*. Cambridge, MA: Harvard University Press, 2011.

Noble, David F. *America by Design*. New York: Oxford University Press, 1977.

Nye, David E. *America as Second Creation*. Cambridge, MA: MIT Press, 2003.

Nye, David E. *American Illuminations: Urban Lighting, 1800–1920*. Cambridge, MA: MIT Press, 2018.

Nye, David E. *American Technological Sublime*. Cambridge, MA: MIT Press, 1994.

Nye, David E. *America's Assembly Line*. Cambridge, MA: MIT Press, 2013.

Nye, David E. *Conflicted American Landscapes*. Cambridge, MA: MIT Press, 2021.

Nye, David E. *Consuming Power: A Social History of American Energies*. Cambridge, MA: MIT Press, 1998.

Nye, David E. *Electrifying America: Social Meanings of a New Technology*. Cambridge, MA: MIT Press, 1990.

Nye, David E. *Henry Ford: Ignorant Idealist*. Port Washington, NY: Kennikat Press, 1979.

Nye, David E. *Image Worlds*. Cambridge, MA: MIT Press, 1985.

Nye, David E. *The Invented Self: An Anti-Biography, from Documents of Thomas A. Edison*. Odense: Odense University Press, 1983.

Nye, David E. "A Model for Heterogeneous Energy Transitions." In *Electrical Conquest: New Approaches to the History of Electrification*, ed. W. Bernard Carlson and Erik M. Conway, 21–50. Cham, Switzerland: Springer, 2023.

Nye, David E. *Narratives and Spaces: Technology and the Construction of American Culture*. New York: Columbia University Press, 1997.

Nye, David E. *Seven Sublimes*. Cambridge, MA: MIT Press, 2022.

Nye, David E. *Technology Matters: Questions to Live With*. Cambridge, MA: MIT Press, 2006.

Oettermann, Stephan. *The Panorama: History of a Mass Medium*. New York: Zone Books, 1997.

Oldenziel, Ruth. *Making Technology Masculine*. Amsterdam: Amsterdam University Press, 1999.

O'Malley, Michael. *Keeping Watch: A History of American Time*. New York: Viking Press, 1990.

Opie, John. *Ogallala: Water for a Dry Land*, 2nd ed. Lincoln: University of Nebraska Press, 2000.

Oppenheim, James. *The Olympian*. New York: Harper, 1912.

Orvell, Miles. *The Death and Life of Main Street: Small Towns in American Memory, Space, and Community*. Chapel Hill: University of North Carolina, 2012.

Orvell, Miles. *The Real Thing: Imitation and Authenticity in American Culture, 1880–1920*. Chapel Hill: University of North Carolina Press, 1989.

O'Shea, P. F. "The Shop Paper as an Aid to Morale." *Factory* 23 (September 1919): 791–793.

Oshima, Harry T. "The Growth of US Factor Productivity." *Journal of Economic History* 44, no. 1 (March 1984): 161–170.

Øverland, Orm. *Immigrant Minds, American Identities*. Urbana: University of Illinois, 2000.

Page, Max. *The City's End: Two Centuries of Fantasies, Fears and Premonitions of New York's Destruction*. New Haven, CT: Yale University Press, 2008.

Paine, Albert Bigelow. *Mark Twain, A Biography*. 2 vols. New York: Harper & Brothers, 1912.

Paine, Albert Bigelow. "The New Coney Island." *Century Magazine* 68 (1905): 528–538.

Palmer, A. Mitchell. "The Case Against the Reds." *The Forum* 63 (February 1920): 63–75.

Palmer, Christopher. *Castaway Tales: From "Robinson Crusoe" to "Life of Pi."* Middletown, CT: Wesleyan University Press, 2016.

Paul, Heike. *The Myths That Made America: An Introduction to American Studies*. Bielefeld: Transcript Verlag, 2014.

Peterson, Robert. *Only the Ball Was White: A History of Legendary Black Players and All-Black Professional Teams*. New York: Oxford University Press, 1992.

Pivar, David J. *Purity Crusade: Sexual Morality and Social Control, 1868–1900*. Westport, CT: Greenwood Press, 1973.

Platt, Harold. *The Electric City: Energy and the Growth of the Chicago Area, 1880–1930*. Chicago: University of Chicago Press, 1991.

Plunkitt, George Washington. "The Curse of Civil Service Reform." In *Plunkitt of Tammany Hall*, ed. William L. Riordon. New York: Penguin, [1903] 2015. Also available at https://glc.yale.edu/curse-civil-service-reform.

Popp, Richard. "Machine Age Communication: Media, Transportation, and Contact in the Interwar United States." *Technology and Culture* 52, no. 3 (2011), 459-484.

"The President's Message." *The Outlook*, May 12, 1906.

Raban, Jonathan. *Bad Land*. New York: Random House, 1997.

Rabenbach, Anson. *The Human Motor: Energy, Fatigue, and the Origins of Modernity*. New York: Basic Books, 1990.

Rabenek, Andrew. "Edison and Modern Construction: The Longue Durée of the Long Kiln." In *Water, Doors and Buildings: The Proceedings of the Sixth Conference of the Construction History Society*, ed. James Campbell et al., 463–477. Cambridge: Queen's College, 2019.

Register, Woody. *The Kid of Coney Island*. New York: Oxford University Press, 2001.

Reik, Henry Ottridge. *A Tour of America's National Parks*. New York: E. P. Dutton, 1927.

Reynolds, Terry S. "The Education of Engineers in America Before the Morrill Act of 1862." *History of Education Quarterly* 32, no. 4 (Winter 1992): 459–482.

Richards, Claude. *The Man of Tomorrow: A Discussion of Vocational Success with the Boy of Today*. New York: Thomas T. Crowell, 1917.

Riis, Jacob. *How the Other Half Lives*. New York: Charles Scribner's Sons, 1890.

Ritchie, Andrew. *Major Taylor: The Extraordinary Career of a Champion Bicycle Racer*. Baltimore: Johns Hopkins University Press, 1988.

Robertson, James Oliver. *America's Business*. New York: Hill and Wang, 1985.

Rodgers, Daniel T. *The Work Ethic in Industrial America, 1850–1920*. Chicago: University of Chicago Press, 1978.

Roosevelt, Theodore. "The Present State of Civil Service Reform." *The Atlantic Monthly*, February 1895, 238–246. https://www.theatlantic.com/magazine/archive/1895/02/the-present-status-of-civil-service-reform/519723/.

Roosevelt, Theodore. *The Winning of the West*. 6 vols. New York: G. P. Putnam's, 1900.

Rosenberg, Nathan, and Manuel Trajtenberg. "A General-Purpose Technology at Work: The Corliss Steam Engine in the Late-Nineteenth-Century United States." *Journal of Economic History* 64, no. 1 (March 2004): 66–99.

Runte, Alfred. *National Parks: The American Experience*. Lincoln: University of Nebraska Press, 1987.

Rydell, Robert. *All the World's a Fair*. Chicago: University of Chicago Press, 1984.

Samuelson, Paul. "An Exact Consumption-Loan Model of Interest with or Without the Social Contrivance of Money." *Journal of Political Economy* 66, no. 6 (1958): 467–482.

Sandberg, Carl. *The Chicago Race Riots: July 1919*. New York: Harcourt, Brace and Howe, 1919.

Schatzberg, Eric. *Technology: Critical History of a Concept*. Chicago: University of Chicago Press, 2018.

Schiller, Dan. *Objectivity and the News: The Public and the Rise of Commercial Journalism*. Philadelphia: University of Pennsylvania Press, 1981.

Schleier, Merrill. *The Skyscraper in American Art*. New York: Da Capo, 1986.

Schlereth, Thomas J. *Victorian America: Transformations in Everyday Life, 1876–1915*. New York: Harper-Perennial, 1991.

Schneirov, Richard. "Thoughts on Periodizing the Gilded Age: Capital Accumulation, Society, and Politics, 1873–1898." *Journal of the Gilded Age and Progressive Era* 5, no. 3 (July 2006): 189–224.

Schudson, Michael. *Advertising, the Uneasy Persuasion: Its Dubious Impact on American Society*. New York: Basic Books, 1984.

Schutz, Alfred, and Thomas Luckmann. *The Structures of the Lifeworld*. Evanston, IL: Northwestern University Press, 1973.

Scranton, Philip. *Endless Novelty: Specialty Production and American Industrialization, 1865–1925*. Princeton, NJ: Princeton University Press, 1997.

Scranton, Philip, and Patrick Fridenson. *Reimagining Business History*. Baltimore: Johns Hopkins University Press, 2013.

Sears, John. *Sacred Places: American Tourist Attractions in the Nineteenth Century*. New York: Oxford University Press, 1989.

Sears, Roebuck Corporation. *The 1902 Edition of the Sears, Roebuck Catalogue*. New York: Bounty Books, [1969] 1986.

Seely, Bruce E. *Building the American Highway System: Engineers as Policy Makers*. Philadelphia: Temple University Press, 1987.

Selcraig, Bruce. "The Real Robinson Crusoe." *Smithsonian Magazine*, July 2005. https://www.smithsonianmag.com/history/the-real-robinson-crusoe-74877644/.

Selig, Diana. *The Cultural Gifts Movement*. Cambridge, MA: Harvard University Press, 2008.

Shackleton, Robert. "Total Factor Productivity Growth in Historical Perspective." Congressional Budget Office, 2013. https://www.cbo.gov/sites/default/files/113th-congress-2013-2014/workingpaper/44002_TFP_Growth_03-18-2013_1.pdf.

Sharp, George. *City Life and Its Amelioration*, Boston: Richard G. Badger, 1915.

Shaw, Ronald E. *Erie Water West: A History of the Erie Canal, 1794–1854*. Lexington: University of Kentucky Press, 1966.

Sheibach, R. Bruce, et al. "Controlling Acid Mine Drainage from the Picher Mining District." *International Journal of Mine Water* 1 (1982): 45–52.

Shi, David. *The Simple Life*. New York: Oxford University Press, 1985.

Shindo, Charles J. *1927 and the Rise of Modern America*. Lawrence: University Press of Kansas, 2010.

Siegfried, André. *America Comes of Age* (English translation of *Les États-Unis d'aujourd'hui*). New York: Harcourt, Brace, 1927.

Silber, Nina. *The Romance of Reunion*. Chapel Hill: University of North Carolina Press, 1993.

Sinclair, Bruce. *Centennial History of the American Society of Mechanical Engineers*. Toronto: University of Toronto Press, 1980.

Sinclair, Upton. *The Jungle*. New York: Doubleday, Page, 1906.

Sirovich, William. *Congressional Record*, May 25, 1928, 9937–9938.

Sklar, Martin. *The Corporate Reconstruction of American Capitalism*. Cambridge: Cambridge University Press, 1988.

Skrabec, Quentin R. *William McKinley, Apostle of Protectionism*. New York: Algora, 2007.

Smil, Vaclav. *Energy and Civilization*. Cambridge, MA: MIT Press, 2017.

Smil, Vaclav. *Energy Transitions: History, Requirements, Prospects*. New York: Praeger, 2010.

Smil, Vaclav. *Energy in World History*. Boulder, CO: Westview, 1994.

Smith, George. "The Survey's Contributions to the National Parks Movement." In *Proceedings of the National Parks Conference*. Washington, DC: US Government Printing Office, 1917.

Smith, Jacob. *Vocal Tracks: Performance and Sound Media*. Berkeley: University of California Press, 2008.

Smith, Matthew Hale. *Sunshine and Shadow in New York*. Hartford: J. B. Burr, 1869.

Smith, Merritt Roe. *Harpers Ferry Armory and the New Technology*. Ithaca, NY: Cornell University Press, 1977.

Smith, Merritt Roe, and Leo Marx. *Does Technology Drive History? The Dilemma of Technological Determinism*. Cambridge, MA: MIT Press, 1994.

Smith, Zachary. *Age of Fear: Othering and American Identity During World War I*. Baltimore: Johns Hopkins University Press, 2019.

Smyth, William E. *The Conquest of Arid America*. New York: Macmillan, 1905.

Sombart, Werner. *Why Is There No Socialism in the United States?* London: Routledge, [1906] 1977.

Sovacool, Benjamin K. "How Long Will It Take? Conceptualizing the Temporal Dynamics of Energy Transitions." *Energy Research and Social Science* 13 (2016): 202–215.

Spence, Mark David. *Dispossessing the Wilderness: Indian Removal and the Making of the National Parks*. New York: Oxford University Press, 2000.

Stead, W. T. *The Americanization of the World; or, The Trend of the Twentieth Century*. London: H. Markley, 1902.

Steffens, Lincoln. *The Shame of the Cities*. New York: McClure, Phillips, 1904.

Steinberg, Ted. *Down to Earth: Nature's Role in American History*. New York: Oxford University Press, 2002.

"Stenographic Report of the Proceedings of Light's Golden Jubilee." The Henry Ford, Greenfield Village, Dearborn, 1929.

Sterling, Christopher H., and John M. Kittross. *Stay Tuned*. Belmont, CA: Wadsworth, 1978.

Stilgoe, John. *Metropolitan Corridor*. New Haven, CT: Yale University Press, 1983.

Stover, John F. *The Routledge Historical Atlas of the American Railroads*. New York: Routledge, 1999.

Strong, Josiah. *Our Country*. New York: American Home Missionary Society, 1885.

Stuckey, Sterling. *Slave Culture: Nationalist Theory and the Foundations of Black America*. Oxford University Press, 1987.

Sydnor, Synthia. "A History of Synchronized Swimming." *Journal of Sport History* 25, no. 2 (1998): 252–267.

Tarbell, Ida. *The History of Standard Oil*. New York: McClure, Phillips, 1904.

Tattersall, Mason. "Thermal Degeneration: Thermodynamics and the Heat-Death of the Universe in Victorian Science, Philosophy, and Culture." In *Decadence, Degeneration, and the End*, ed. Marja Härmänmaa and Christopher Nissen. New York: Palgrave Macmillan, 2014.

Taylor, Benjamin F. *Between the Gates*. Chicago: S. C. Griggs, 1878.

Taylor, Carol M. "W. E. B. Du Bois's Challenge to Scientific Racism." *Journal of Black Studies* 11, no. 4 (June 1981): 449–460.

Taylor, Frederick Winslow. *The Principles of Scientific Management*. New York: Harper & Brothers, 1911.

Taylor, George Rogers. *The Transportation Revolution*. New York: Harper & Row, 1951.

Thayer, William M. *From Log Cabin to White House*. Boston: James H. Earle, 1881.

"Thomas A. Edison, Guest of the Institute." *Western Electrician*, February 20, 1904, 151–153.

Thoreau, Henry David. *Walden*. Boston: Ticknor & Fields, 1854.

Thurston, Robert H. "The Trend of National Progress." *North American Review* 161 (September 1895): 297–313.

Tichi, Cecelia. *Shifting Gears: Technology, Literature, Culture in Modernist America*. Chapel Hill: University of North Carolina Press, 1987.

Tierney, John. "What Is Nostalgia Good For? Quite a Bit, Research Shows." *New York Times*, July 8, 2013.

Tocqueville, Alexis de. *Democracy in America*, 2 vols. New York: Vintage Books, 1945.

Trachtenberg, Alan. *The Incorporation of America*. New York: Hill and Wang, 1982.

Trachtenberg, Alan. *Reading American Photographs*. New York: Hill and Wang, 1989.

Trescott, Martha Moore. "The Bicycle, a Technical Precursor of the Automobile." *Business and Economic History* 5 (1976): 51–75.

Trine, Ralph Waldo, and Henry Ford. *The Power That Wins*. Indianapolis: Bobbs Merrill, 1929.

Tsujimoto, Karen. *Images of America: Precisionist Painting and Modern Photography*. Seattle: University of Washington Press, 1982.

Turner, Frederick Jackson. *The Frontier in American History*. New York: Henry Holt, 1920.

Twain, Mark. *A Connecticut Yankee in King Arthur's Court*. New York: Charles L. Webster, 1889.

Twain, Mark. *Roughing It*. New York: Harper & Brothers, 1913.

Twain, Mark, and Charles Dudley Warner. *The Gilded Age*. Hartford, CT: American Publishing, 1874.

Upward, Geoffrey C. *A Home for Our Heritage: The Building and Growth of Greenfield Village and Henry Ford Museum, 1929–1979*. Dearborn, MI: Edison Institute, 1979.

Valosik, Vicki. "Synchronized Swimming Has a History That Dates Back to Ancient Rome." *Smithsonian Magazine*, August 12, 2016.

Van Slyck, Abigail A. *A Manufactured Wilderness: Summer Camps and the Shaping of American Youth, 1890–1960*. Minneapolis: University of Minnesota Press, 2010.

Vella, Christine. *George Washington Carver*. Baton Rouge: Louisiana State University Press, 2015.

Wallace, Jerry L. *Calvin Coolidge, Our First Radio President*. Plymouth Notch, VT: Calvin Coolidge Memorial Foundation, 2008.

Ward, John William. "The Meaning of Lindbergh's Flight." *American Quarterly* 10, no. 1 (Spring 1958): 3–16.

Watney, Simon. "Making Strange: The Shattered Mirror." In *Thinking Photography*, ed. Victor Burgin. London: Macmillan, 1982.

Wattenberg, Ben J. *The Statistical History of the United States: From Colonial Times to the Present*. New York: Basic Books, 1976.

Webb, Sheila. "A Pictorial Myth in the Pages of *Life*: Small-Town America as the Ideal Place." *Studies in Popular Culture* 28, no. 3 (2006): 35–58.

Wells, H. G. *The Time Machine*. London: William Heineman, 1895.

Whitaker, Jan. "Before Horn & Hardart: European Automats." https://restaurant-ingthroughhistory.com/2013/04/30/before-horn-hardart-european-automats-2/.

White, Eugene N. "The Stock Market Boom and Crash of 1929 Revisited." *Journal of Economic Perspectives* 4, no. 2 (1990): 67–83.

White, Michael V. "Reading and Rewriting: The Production of an Economic Robinson Crusoe." *Southern Review*, July 1982, 115–142.

White, Richard. *Railroaded: The Transcontinentals and the Making of Modern America.* New York: W. W. Norton, 2011.

Wiebe, Robert. *The Search for Order, 1877–1920.* New York: Hill and Wang, 1967.

Williams, Rosalind. *Dream Worlds: Mass Consumption in Late Nineteenth-Century France.* Berkeley: University of California Press, 1982.

Williams, William Carlos. *Sour Grapes.* Boston: Four Seas, 1921.

Wilson, Richard Guy, Dianne H. Pilgrim, and Dickran Tashjian. *The Machine Age in America, 1918–1941.* New York: Harry N. Abrams, 2001.

Wilson, William H. *The City Beautiful Movement.* Baltimore: Johns Hopkins University Press, 1989.

Wister, Owen. *The Virginian.* New York: Macmillan, 1902.

Wohl, Robert. *A Passion for Wings: Aviation and the Western Imagination, 1908–1918.* New Haven, CT: Yale University Press, 1994.

Wosk, Julie H. "The Impact of Technology on the Human Image in Art." *Leonardo* 19, no. 2 (April 1986): 145–152.

Wright, Harold Bell. *The Winning of Barbara Worth.* New York: A. L. Burt, 1911.

Wright, Rebecca. *Moral Energy in America.* Baltimore: Johns Hopkins University Press, 2025.

Yost, Casper S. *The Principles of Journalism.* New York: Appleton, 1924.

Zeller, Thomas. *Consuming Landscapes: What We See When We Drive and Why It Matters.* Baltimore: Johns Hopkins University Press, 2022.

Zinn, Howard. *A People's History of the United Sates.* London: Pearson, 2003.

INDEX

Publisher contact:
The MIT Press
Massachusetts Institute of Technology
77 Massachusetts Avenue, Cambridge, MA 02139
mitpress.mit.edu

EU Authorised Representative:
Easy Access System Europe, Mustamäe tee 50,
10621 Tallinn, Estonia
gpsr.requests@easproject.com

Printed by Integrated Books International,
United States of America